NIST Special Publication 800-65

National Institute of Standards and Technology
Technology Administration
U.S. Department of Commerce

Integrating IT Security into the Capital Planning and Investment Control Process

Joan Hash, Nadya Bartol, Holly Rollins, Will Robinson, John Abeles, and Steve Batdorff

INFORMATION SECURITY

Computer Security Division
Information Technology Laboratory
National Institute of Standards and Technology
Gaithersburg, MD 20899-8930

Version 1.0

January 2005

Reports on Information Systems Technology

The Information Technology Laboratory (ITL) at the National Institute of Standards and Technology promotes the United States economy and public welfare by providing technical leadership for the Nation's measurement and standards infrastructure. ITL develops tests, test methods, reference data, proof-of-concept implementations, and technical analyses to advance the development and productive use of information technology. ITL's responsibilities include the development of management, administrative, technical, and physical standards and guidelines for the cost-effective security and privacy of non-national-security-related information in federal information systems. This Special Publication 800 series reports on ITL's research, guidelines, and outreach efforts in information system security and its collaborative activities with industry, government, and academic organizations.

Authority

This document has been developed by the National Institute of Standards and Technology (NIST) to further respond to its statutory responsibilities under the Federal Information Security Management Act of 2002, Public Law 107-347.

NIST is responsible for developing standards and guidelines, including minimum requirements, for providing adequate information security for all agency operations and assets, but such standards and guidelines shall not apply to national security systems. This guideline is consistent with the requirements of the Office of Management and Budget (OMB) Circular A-130, Section 8b(3), Securing Agency Information Systems, as analyzed in A-130, Appendix IV: Analysis of Key Sections. Supplemental information is provided in A-130, Appendix III.

This guideline has been prepared for use by federal agencies. It may be used by nongovernmental organizations on a voluntary basis and is not subject to copyright. (Attribution would be appreciated by NIST.)

Nothing in this document should be taken to contradict standards and guidelines made mandatory and binding on federal agencies by the Secretary of Commerce under statutory authority. Nor should these guidelines be interpreted as altering or superseding the existing authorities of the Secretary of Commerce, Director of the OMB, or any other federal official.

> Certain commercial entities, equipment, or materials may be identified in this document in order to describe an experimental procedure or concept adequately. Such identification is not intended to imply recommendation or endorsement by the National Institute of Standards and Technology, nor is it intended to imply that the entities, materials, or equipment are necessarily the best available for the purpose.

Acknowledgements

The authors would like to thank all of those who assisted in reviewing early drafts of this guidance as well as those who participated in related workshops and meetings and provided their comments to assist in developing this special publication. Also, the authors would like to thank the diligent editorial reviews by Linda Duncan and Elizabeth Lennon and the graphics support provided by Sergei Kalinin.

TABLE OF CONTENTS

Executive Summary ES-1

1. Introduction 1
   1.1 Background 1
   1.2 Relationship to Existing Guidance 1
   1.3 Purpose and Scope 2
   1.4 Security CPIC Process Overview 3
   1.5 Definitions 4
   1.6 Audience 5
   1.7 Document Organization 5

2. Legislative and Regulatory Environment Overview 6
   2.1 Reporting Requirements 6
   2.2 Select-Control-Evaluate Investment Life Cycle 8
       2.2.1 Select Phase 9
       2.2.2 Control Phase 10
       2.2.3 Evaluate Phase 11
   2.3 Earned Value Management 11
   2.4 Information Technology Investment Management 12
   2.5 Plan of Action and Milestones 13
   2.6 Risk 15

3. IT Security and Capital Planning Integration Roles and Responsibilities 18
   3.1 Overview 19
   3.2 Head of Agency 19
   3.3 Senior Agency Officials 20
   3.4 Chief Information Officer 20
   3.5 Senior Agency Information Security Officer 21
   3.6 Chief Financial Officer 21
   3.7 Investment Review Board 21
   3.8 Technical Review Board 21
   3.9 IT Capital Planning, Architecture, and Security and Privacy Subcommittees ... 22
   3.10 Operating Unit/Bureau Executive Management 22
   3.11 Project Manager 22
   3.12 System Owner 22

4. Integration of Security into the CPIC Process 23
   4.1 The CPIC Process and IT Security 23
   4.2 Identify Baseline 25
   4.3 Identify Prioritization Criteria 26
   4.4 Prioritize Against Requirements 28
       4.4.1 Enterprise-level Prioritization 29
       4.4.2 System-Level Prioritization 33
       4.4.3 Joint Prioritization 35
   4.5 Develop Supporting Materials 37
       4.5.1 Enterprise- and System-Level Considerations 37
       4.5.2 Concept Paper 37
       4.5.3 Investment Thresholds 38
       4.5.4 The Exhibit 300 38
   4.6 IRB and Portfolio Management 41
   4.7 Exhibits 53, 300, and Program Management 41

5. Implementation Issues 43
    5.1   IT Security Organizational Processes 43
    5.2   Project Management 44
    5.3   Legacy Systems 44
    5.4   Timelines 45

Appendix A. Glossary 47

Appendix B. Acronyms 50

Appendix C. References 51

Appendix D. Security Requirements Mapping 52

LIST OF FIGURES

Figure ES-1. Federal IT Security and Capital Planning Legislation, Regulations, and Guidance ES-1

Figure ES-2. Integrating IT Security and Capital Planning ES-3

Figure 1-1. NIST Special Publications and the Prioritization Process 2

Figure 1-2. Integrating IT Security and Capital Planning 3

Figure 2-1. Federal IT Security and Capital Planning Legislation, Regulations, and Guidance 6

Figure 2-2. The Select-Control-Evaluate Investment Life Cycle 9

Figure 2-3. ITIM Maturity Model 12

Figure 2-4. POA&M Quarterly Update Format 15

Figure 3-1. Notional IT Management Hierarchy 18

Figure 3-2. Roles and Responsibilities Throughout the CPIC Process 19

Figure 4-1. Integrating IT Security Into the CPIC Process 23

Figure 4-2. IT Security Activities Throughout the Investment Life Cycle 24

Figure 4-3. Investment Life Cycle and SDLC Decision-Making 25

Figure 4-4. Identifying Baseline Best Practices 26

Figure 4-5. Corrective Action Impact 29

Figure 4-6. Enterprise-Level Prioritization 31

Figure 4-7. Enterprise-Level Prioritization Analysis Matrix 32

Figure 4-8. System-Level Prioritization 34

Figure 4-9. System-level Prioritization Analysis Matrix 35

Figure 4-10. Joint Prioritization Analysis Matrix 35

Figure 4-11. Corrective Action Prioritization with Costs 36

Figure 4-12. Enterprise- and System-Level Requirements 37

Figure 4-13. Illustrative Project Thresholds 38

Figure 4-14. The Investment Process, Culminating in the Exhibit 300 39

Figure 5-1. Layers of Integration of Security into the CPIC Process 43

Figure 5-2. Budget Timelines 45

Figure 5-3. CPIC Timelines 46

LIST OF TABLES

Table 2-1. POA&M Sections 14
Table 4-1. Exhibit 300 Requirements 40
Table D-1. Security Requirements Mapping 52

Executive Summary

Traditionally, information technology (IT) security and capital planning and investment control (CPIC) processes have been performed independently by security and capital planning practitioners. However, the Federal Information Security Management Act (FISMA) of 2002 and other existing federal regulations charge agencies with integrating the two activities. In addition, with increased competition for limited federal budgets and resources, agencies must ensure that available funding is applied towards the agencies' highest priority IT security investments. Applying funding towards high-priority security investments supports the objective of maintaining appropriate security controls, both at the enterprise-wide and system level, commensurate with levels of risk and data sensitivity. This special publication (SP) introduces common criteria against which agencies can prioritize security activities to ensure that corrective actions identified in the annual FISMA reporting process are incorporated into the capital planning process to deliver maximum security in a cost-effective manner.

The implementation of IT security and capital planning practices within the federal government is driven by a combination of legislation, rules and regulations, and agency-specific policies. FISMA requires agencies to integrate IT security into their capital planning and enterprise architecture processes, conduct annual IT security reviews of all programs and systems, and report the results of those reviews to the Office of Management and Budget (OMB). Therefore, the implementation of FISMA legislation effectively integrates IT security and capital planning because agencies must document resource and funding plans for IT security. Furthermore, implementation of FISMA legislation is intended to ensure that agency resources are protected and risk is effectively managed. It requires that agencies incorporate IT security into the life cycle of their information systems. OMB's FISMA reporting guidance also referenced use the National Institute of Standards and Technology (NIST) SP 800-26, *Security Self-Assessment Guide for Information Technology Systems* to evaluate agency security programs. The results of the self-assessment should be documented in the agency's annual FISMA report and logged in the agency's plan of action and milestones (POA&M), along with POA&M inputs from other appropriate sources. The agency must then determine the costs and timeframes associated with mitigating the weaknesses identified in the POA&Ms. These costs are captured in the system or program's annual OMB Exhibit 300 and in the enterprise-wide Exhibit 53, which are the funding vehicles submitted to OMB to secure an operating budget. Figure ES-1 illustrates this process.

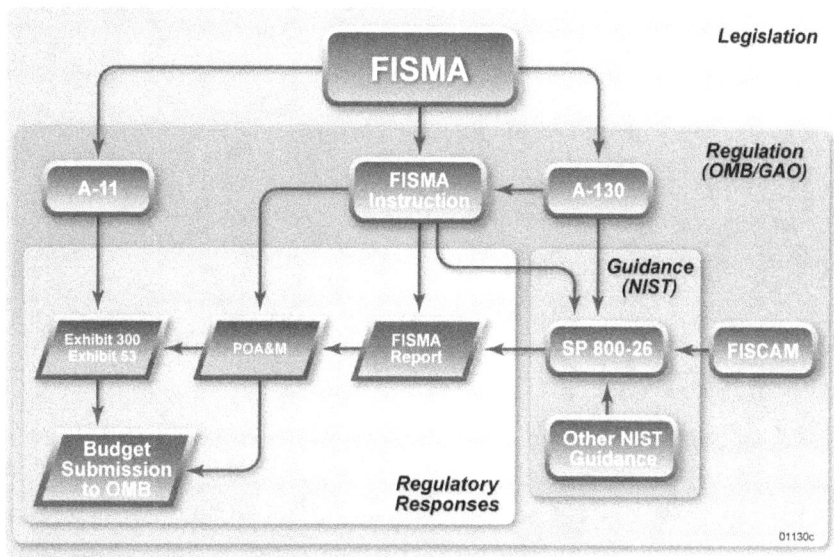

Figure ES-1. Federal IT Security and Capital Planning Legislation, Regulations, and Guidance

The investment life cycle employed by agencies for IT investments is based on the Government Accountability Office's (GAO) Select, Control, and Evaluate framework. This investment management framework ensures that investment management practices, including IT security, are disciplined and thorough throughout the investment's life cycle.

To address the capital planning and IT security requirements imposed on federal IT investments, NIST recommends a seven-step framework (see Figure ES-2) for integrating IT security into the capital planning process for enterprise-level IT security activities and individual system IT security activities:

- **Enterprise-level investments** – those security investments that are ubiquitous across the agency and will improve the overall agency's security posture. [For example, an enterprise-wide firewall or intrusion detection system (IDS) acquisition or public key infrastructure (PKI).]

- **System-level investments** – those security investments designed to strengthen a discrete system's security posture. (For example, strengthening password controls or testing a contingency plan for a particular system.)

The framework assists federal agencies in integrating IT security into the capital planning process by providing a systematic approach to selecting, managing, and evaluating IT security investments. The methodology relies on existing data inputs (for example, NIST SP 800-26 self-assessment, certification and accreditation information, and audit reports) so it can be readily implemented at federal agencies. Inputs for the methodology include:

- **Enterprise-Level Information**
    - Stakeholder rankings of enterprise-wide initiatives
    - Enterprise-wide initiative IT security status
    - Cost of implementing remaining appropriate security controls for enterprise-wide initiatives

- **System-Level Information**
    - System categorization[1]
    - Security compliance
    - Corrective action cost

---

[1] See NIST Federal Information Processing Standard (FIPS) 199, *Standard for Security Categorization of Federal Information and Information Systems*

Based on the above information, the seven-step methodology, shown in Figure ES-2, can help agencies identify high-priority corrective actions for immediate funding. The seven steps include:

1. **Identify the Baseline:** use information security metrics or other available data to baseline the current security posture.

2. **Identify Prioritization Requirements:** evaluate security posture against legislative and Chief Information Officer (CIO)-articulated requirements and agency mission.

3. **Conduct Enterprise-Level Prioritization:** prioritize potential enterprise-level IT security investments against mission and financial impact of implementing appropriate security controls.

4. **Conduct System-Level Prioritization:** prioritize potential system-level corrective actions against system category and corrective action impact.

5. **Develop Supporting Materials:** for enterprise-level investments, develop concept paper, business case analysis, and Exhibit 300. For system-level investments, adjust Exhibit 300 to request additional funding to mitigate prioritized weaknesses.

6. **Implement Investment Review Board (IRB) and Portfolio Management:** prioritize agency-wide business cases against requirements and CIO priorities and determine investment portfolio.

7. **Submit Exhibit 300s, Exhibit 53, and Conduct Program Management:** ensure approved 300s become part of the agency's Exhibit 53; ensure investments are managed through their life cycle (using Earned Value Management for Development/Modernization/Enhancement investments and operational assessments for steady state investments) and through the GAO's Information Technology Investment Management (ITIM) maturity framework.

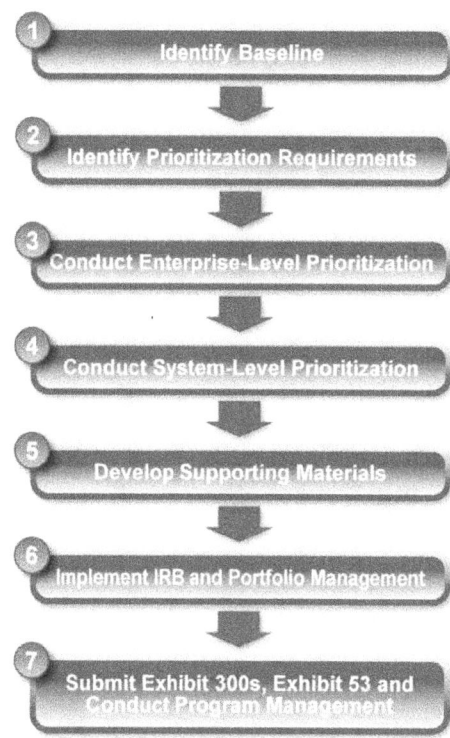

Figure ES-2. Integrating IT Security and Capital Planning

The process presented in this guidance is intended to serve as a model methodology. Agencies should work within their investment planning environments to adapt and incorporate the pieces of this process into their own unique processes to develop workable approaches for CPIC. If incorporated into an agency's processes, the methodology can help ensure that IT security is appropriately planned for and funded throughout the investment's life cycle, thus strengthening the agency's overall security posture.

1. Introduction

1.1 Background

With the release of the Federal Information Security Management Act (FISMA) in 2002, the need for information technology (IT) security guidance within the federal community has increased. Capital planning was once seen as applying primarily to IT systems. With FISMA underscoring the emphasis on IT security at both the system and enterprise levels, security investments must now be brought into the capital planning process. FISMA, the Clinger-Cohen Act, and other associated guidance and regulations, including Office of Management and Budget (OMB) Circulars A-11 and A-130, charge agencies with integrating IT security and the capital planning and investment control (CPIC) process.

> This special publication was developed under the assumption that the reader possesses a basic familiarity with requisite IT security and capital planning guidance and legislation including FISMA, OMB Circulars A-11 and A-130, the Clinger-Cohen Act, NIST special publications, and is familiar with IT security controls and requirements. While detailed knowledge of these regulations and guidance documents is not essential to understanding this special publication, a basic familiarity with these regulations and guidance would assist with comprehension of this document.

Determining the benefit to the agency from IT security investments is a key criterion of IT security planning. Traditionally, IT security and capital planning have been thought of as separate activities by security and capital planning practitioners. However, with FISMA legislation and existing federal regulations that charge agencies with integrating the two activities and with increased competition for limited federal budgets, agencies must effectively integrate their IT security and capital planning processes. This guidance introduces common criteria against which agencies can prioritize security activities and ensure that corrective actions identified during the FISMA reporting process are incorporated into the capital planning process to deliver maximum security and financial benefit to the agency.

The National Institute of Standards and Technology (NIST) first explored this topic in its 2002–2003 Return on Security Investment (ROSI) study. During this effort, NIST interviewed Chief Information Officers (CIO), Chief Financial Officers (CFO), and Chief Technology Officers of federal agencies and private sector companies to generate a common body of knowledge and to identify best practices in returns on IT security investments in both the public and private sectors.

NIST used the information collected through the ROSI study as the foundation for a workshop on integrating security and capital planning efforts. On June 4, 2003, and June 30, 2003, NIST presented a workshop entitled *Integrating IT Security into the Capital Planning and Investment Control Process*. Over 200 members of the federal community attended the two workshops, where they learned how to prioritize security investments to ensure that the most cost-effective, highest impact investments would receive funding. This document captures and expands upon the proceedings of the two workshops, including the prioritization process.

1.2 Relationship to Existing Guidance

This document is a continuation in a series of NIST special publications (SP) intended to assist IT security personnel in planning and prioritizing their IT security investments. This document illustrates a prioritization approach that uses the 17 topic areas found in NIST SP 800-26, *Security Self-Assessment Guide for Information Technology Systems*, as investment prioritization criteria for multiple types of investments, including plan of actions and milestones (POA&M) corrective actions. While other criteria may be substituted for these topic areas (for example, recommended security controls from NIST SP 800-53, *Recommended Security Controls for Federal Information Systems*), using NIST SP 800-26 was deemed favorable because it enables agencies to reuse data and information they have already gathered to support FISMA reporting, thus substantially reducing the need for additional data collection. Figure 1-1 illustrates the relationship between NIST SP 800-65 and other NIST guidance.

In addition to NIST SP 800-26, the security metrics development and implementation methodology provided in NIST SP 800-55, *Security Metrics Guide for Information Technology Systems*, can be used as

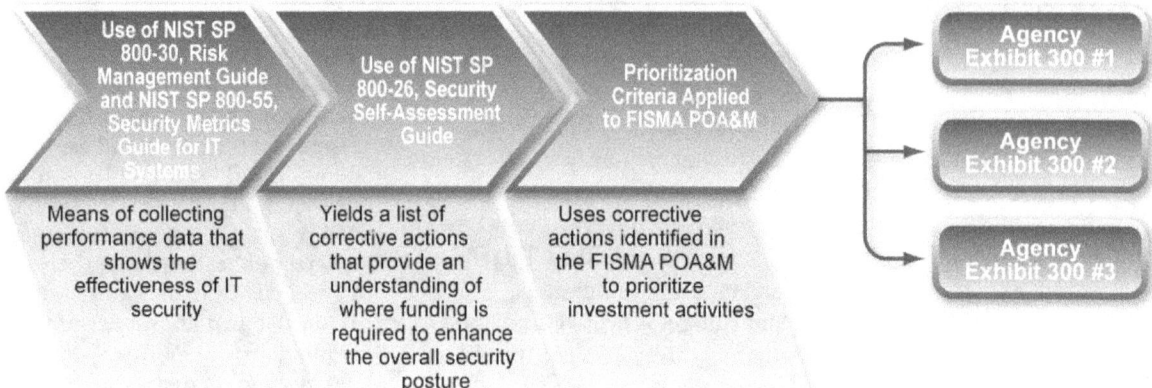

Figure 1-1. NIST Special Publications and the Prioritization Process

a source for baselining an agency's IT security posture, and NIST SP 800-30, *Risk Management Guide for Information Technology Systems*, can be used to identify vulnerabilities before completing an OMB Exhibit 300.

NIST SP 800-30 provides definitions and practical guidance necessary for assessing and mitigating security risks identified within IT systems.[2] NIST 800-55 provides an approach for identifying, formulating, and implementing metrics to assess the level of security policies and procedures implementation, gauge efficiency and effectiveness of security measures, and determine the impact of such measures on the organization's mission and business. NIST 800-26 provides a vehicle for an internal assessment of the controls in place for an application or a general support system.[3] All three pieces of guidance define a set of activities, the results of which provide inputs into the POA&M, and are then prioritized to ensure the most pressing security needs are addressed first. It should be noted that results of other activities, such as weaknesses identified by Government Accountability Office (GAO) and Inspector General (IG) audits, should also be documented in the POA&M. Exhibit 300s that include funding for prioritized corrective actions and other costs of security are developed and submitted to OMB to secure funding either as stand-alone investment requests or as the security component of an IT investment proposal. The benefits of this approach are the traceability across each step and the ability to reuse collected data. In addition, the benefits enable seamless risk management as corrective actions identified through risk assessments, self-assessments, and metrics data collections are documented and used in OMB funding requests.

This document contains several references to OMB guidance, including OMB Circular A-11. This OMB guidance intends to provide notional strategies for providing security inputs to the capital planning process. **This guidance does not supersede Circular A-11; rather, it provides additional information to assist agencies with successfully integrating security into their capital planning processes**.

1.3 Purpose and Scope

This document can be used to assist federal agencies in integrating IT security into their CPIC processes by providing a systematic approach to selecting, managing, and evaluating IT security investments. This approach will support alignment with the Federal Enterprise Architecture (FEA)[4] and will provide a

---

[2] NIST SP 800-30, *Risk Management Guide for Information Technology Systems*, page 1.
[3] NIST SP 800-26, *Security Self-Assessment Guide for Information Technology Systems*, page 4.
[4] The FEA is a collection of interrelated "reference models" designed to facilitate cross-agency analysis and identify duplicative investments, gaps, and opportunities for collaboration within and across federal agencies.

balanced process to support prudent portfolio management. Specifically, the systematic approach can help agencies:

- Identify relevant OMB and other guidance that applies to governing federal government IT security investment decisions
- Explain how current security requirements relate and support the IT CPIC process
- Understand the IT investment management process phases—Select, Control, and Evaluate—as they relate to security investments
- Identify CPIC-related roles and responsibilities required to manage IT security investments
- Explain the best practices IT security management process and why it is important for making sound IT security investment decisions
- Understand how to develop security requirements and appropriate supporting documentation for IT acquisition
- Identify steps and materials required to complete a sound business case in support of investment requests
- Understand implementation issues associated with incorporating IT security into the CPIC process.

The process identified in this document is not a rigid methodology to be followed meticulously by all agencies; rather, it is a roadmap that serves to highlight key activities and practices that are essential to a disciplined approach to security capital planning. The process shows the data required to link FISMA and POA&M corrective actions to capital planning.

1.4 Security CPIC Process Overview

A mature CPIC process is essential for effective investment management within any organization. Figure 1-2 presents a model CPIC approach that integrates IT security considerations.

The process is a roadmap that highlights key capital planning activities. Key events within each step include:

- Identify the Baseline: use information security metrics or other available data to baseline the current security posture.

- Identify Prioritization Requirements: evaluate security posture against legislative and Chief Information Officer (CIO)-articulated requirements and agency mission.

- Conduct Enterprise-Level Prioritization: prioritize potential enterprise-level IT security investments against mission and financial impact of implementing appropriate security controls.

- Conduct System-Level Prioritization: prioritize potential system-level corrective actions against system sensitivity and corrective action impact.

Figure 1-2. Integrating IT Security and Capital Planning

- Develop Supporting Materials: for enterprise-level investments, develop concept paper, business case analysis, and Exhibit 300. For system-level investments, adjust Exhibit 300 to request additional funding to mitigate prioritized weaknesses.

- Implement Investment Review Board (IRB) and Portfolio Management: prioritize agency-wide business cases against requirements and CIO priorities and determine investment portfolio.

- Submit Exhibit 300s, Exhibit 53, and Conduct Program Management: ensure approved 300s become part of the agency's Exhibit 53; ensure investments are managed through their life cycle (using Earned Value Management for Development/Modernization/Enhancement investments and operational assessments for steady state investments) and through the GAO's Information Technology Investment Management (ITIM) maturity framework.

The process presented in this document is intended to serve as a model methodology. Agencies should work within their investment planning environments to adapt and incorporate the pieces of this process into their own unique processes to develop workable approaches for integrating IT security into the CPIC process.

A detailed description of a recommended CPIC process is provided in Section 4.

1.5 Definitions

Throughout this guidance document, references are made to key terms that are essential to understanding the integration of IT security into the capital planning process. These terms are defined below.

- *Security controls* are the management, operational, and technical controls (*e.g.*, safeguards or countermeasures) prescribed for an information system to protect the confidentiality, integrity, and availability of the system and its information.[5]

- An *IT security investment* is an IT application, service or system that is solely devoted to security. For instance, intrusion detection systems (IDS) and public key infrastructure (PKI) are examples of IT security investments.

- Security risk versus investment risk are two distinctly different measures:
    - **Security risk**. The level of impact on agency operations (including mission, functions, image, or reputation), agency assets, or individuals resulting from the operation of an information system given the potential impact of a threat and the likelihood of that threat occurring.[6]
    - **Investment risk**. Risks associated with the potential inability to achieve overall program objectives within defined cost, schedule, and technical constraints. OMB has defined 19 areas of investment risk, all of which are required to be addressed in the Exhibit 300.

---

[5] NIST SP 800-53, *Recommended Security Controls for Federal Information Systems*.
[6] NIST SP 800-53, *Recommended Security Controls for Federal Information Systems*

- Select-Control-Evaluate[7] is an IT investment management process:
    - **Select**. The goal of the selection phase is to assess and prioritize current and proposed IT projects and then create a portfolio of IT projects. In doing so, this phase helps to ensure that the organization (1) selects those IT projects that will best support mission needs and (2) identifies and analyzes a project's risks and returns before spending a significant amount of project funds. A critical element of this phase is that a group of senior executives makes project selection and prioritization decisions based on a consistent set of decision criteria that compares costs, benefits, risks, and potential returns of the various IT projects.
    - **Control**. The control phase consists of managing investments while monitoring for results. Once the IT projects have been selected, senior executives periodically assess the progress of the projects against their projected cost, scheduled milestones, and expected mission benefits.
    - **Evaluate**. The evaluation phase provides a mechanism for constantly improving the organization's IT investment process. The goal of this phase is to measure, analyze, and record results based on the data collected throughout each phase. Senior executives assess the degree to which each project has met its planned cost and schedule goals and has fulfilled its projected contribution to the organization's mission. The primary tool in this phase is the post-implementation review (PIR), which should be conducted once a project has been completed. PIRs help senior managers assess whether a project's proposed benefits were achieved and also help to refine the IT selection criteria to be used in the future.

1.6 Audience

The audience for this document includes executive management, IT managers and security professionals, security program managers, IRB participants, and other financial and budget personnel.

1.7 Document Organization

The remainder of this document is structured as follows:

- Section 2 describes IT security legislative and regulatory environment.
- Section 3 discusses roles and responsibilities related to integrating IT security into the CPIC process.
- Section 4 describes the integration of IT security into the CPIC process.
- Section 5 discusses issues associated with CPIC implementation as it relates to IT security.
- Appendix A contains a glossary of terms used in the document.
- Appendix B lists acronyms and abbreviations used in this document.
- Appendix C lists references used in the document.
- Appendix D maps OMB A-11 guidance to NIST SP 800-26 topic areas, NIST SP 800-53 security control families, and to other NIST guidance.

---

[7] The Select, Control, and Evaluate framework was produced cooperatively by OMB's Office of Information and Regulatory Affairs and the GAO's Accounting and Information Management Division. Source – OMB's Guidance: Evaluating Information Technology Investments, A Practical Guide, Version 1, Office of Information and Regulatory Affairs, Information Policy and Technology Branch, November 1995.

2. Legislative and Regulatory Environment Overview

Implementation of IT security within the federal government is guided by a combination of legislation, rules and regulations, and agency-specific policies. The majority of the guidance addresses governing and executing IT security activities. However, to be funded, IT investments must demonstrate compliance with all applicable requirements specified in the guidance. A lack of compliance with identified security controls indicates weaknesses in IT security that should be mitigated at the enterprise and system levels. Appropriate IT security controls must be thoroughly planned for throughout the investment life cycle. Costs associated with meeting IT security controls and ensuring effective protection of federal IT resources should be accounted for in the capital planning process.

2.1 Reporting Requirements

Signed into law in 2002, FISMA requires departments and agencies to integrate IT security into the capital planning process. Specifically, FISMA:

- Charges OMB and NIST to develop security standards and identify tolerable security risk levels
- Makes NIST standards compulsory for all agencies; FISMA eliminated an agency's ability to obtain waivers on NIST standards [Federal Information Processing Standards (FIPS)]
- Charges agencies to integrate IT security into capital planning.

Figure 2-1 illustrates the relationship between legislation, regulation, and guidance that exists for IT security and capital planning for the federal government.

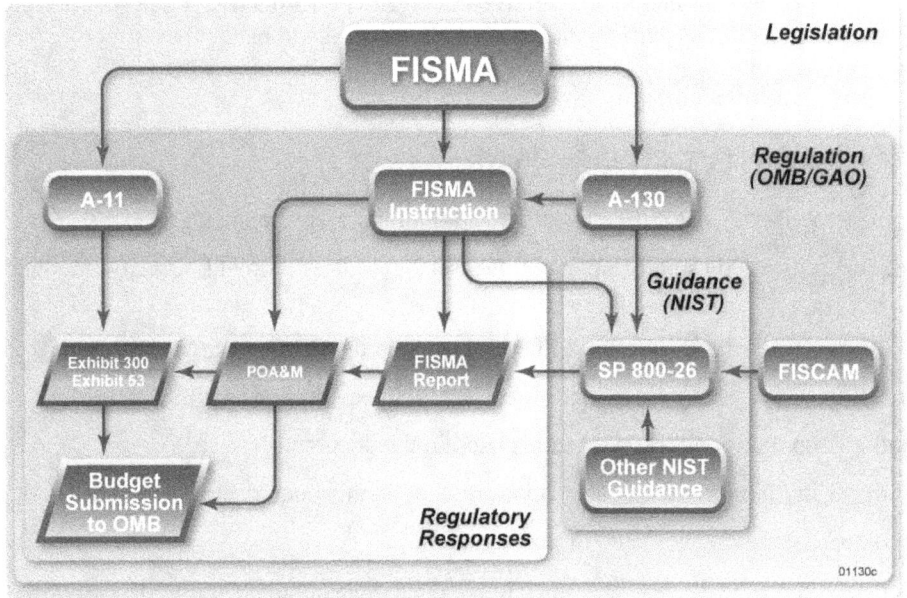

Figure 2-1. Federal IT Security and Capital Planning Legislation, Regulations, and Guidance

FISMA provides overarching requirements for securing federal resources and ensuring that security is incorporated into all phases of the investment life cycle. FISMA codifies specific responsibilities of federal agency officials, addresses protection of agency information resources, calls for agency officials to manage risk to an appropriate level, and requires agencies to incorporate security into the life cycle of information systems. FISMA requires agencies to complete an annual program review that includes conducting self-assessments for all agency systems and conducting a FISMA independent evaluation. Results from these activities are compiled into a comprehensive FISMA report, which is submitted to OMB along with the budget year financial documentation. The corrective actions that agencies identify to mitigate weaknesses found in the FISMA report are documented and tracked in the POA&M.

FISMA reporting includes providing a status of security weaknesses in key areas of a security program. As required by FISMA, OMB provides specific guidance annually. FISMA reporting guidance specifies reporting formats and identifies required actions associated with the quarterly and annual reporting.

NIST SP 800-26 identifies a set of security controls applicable to IT systems and provides a self-assessment checklist that assists program managers and system owners in determining the maturity of their security program implementation. The NIST SP 800-26 self-assessment results are then used by agencies to determine their IT security strengths and vulnerabilities and to provide an overview of the agency's security posture. NIST SP 800-53 defines a set of minimum security controls for information systems in support of the certification and accreditation process.

The capital planning requirements, depicted on the left side of Figure 2-1, illustrate how FISMA and OMB Circular A-11 impact the capital planning process at federal agencies. OMB Circular A-11 directs agencies to complete Exhibit 300s and an Exhibit 53. The Exhibit 300 reflects an investment's plan for capital asset management. In addition, the Exhibit 300 guidance instructs agencies on budget justification and reporting requirements for major acquisitions and major IT systems or projects. Each year, agencies complete and submit an Exhibit 300 for each major IT investment. The Exhibit 300 is an input to the Exhibit 53, which provides the total IT and IT security spending for the year.

The POA&M process provides a direct link to the capital planning process. As Figure 2-1 illustrates, the POA&M information includes the costs of corrective actions that have to be captured in the Exhibit 300 and rolled into the Exhibit 53, which provides an overview of an agency's IT portfolio. The Exhibit 53 includes a rollup of all Exhibit 300s and additional IT expenses from across the agency. All IT investments are identified by mission area and include their budget year and life-cycle cost, as well as the percentage of their costs that are devoted to IT security. All costs are totaled across the agency to provide an overall picture of the agency's IT portfolio.

Costs associated with each POA&M item are required to map to annual budget requests in the Exhibit 300s and the Exhibit 53. These costs are captured as a component of the **percentage of IT security**, or the percentage of the total investment for the budget year associated with IT security in the Exhibit 300, and are then aggregated in the Exhibit 53. Typically, these costs include:[8]

- Direct costs of providing IT security for the specific IT investment. Examples include the following:[9]
    - Risk assessment
    - Security planning and policy
    - Certification and accreditation (C&A)
    - Specific security controls[10]
    - Authentication or cryptographic applications
    - Education, awareness, and training
    - System reviews/evaluations (including system security test and evaluation [ST&E])
    - Oversight or compliance inspections
    - Development or maintenance of agency reports to OMB and corrective action plans as they pertain to the specific investment

---

[8] Appendix D provides a detailed crosswalk from each of these costs to NIST SP 800-26 topic areas and specific NIST guidance documents that describe the implementation of these items.
[9] The list of direct and allocated security costs and benefits was derived from OMB Circular A-11 (2003).
[10] See NIST SP 800-53, *Recommended Security Controls for Federal Information Systems*.

- Contingency planning and testing
- Physical and environmental controls for hardware and software
- Auditing and monitoring
- Computer security investigations and forensics
- Reviews, inspections, audits, and other evaluations performed on contractor facilities and operations
- Privacy impact assessments

- Products, procedures, and personnel that have an incidental or integral component and/or a quantifiable benefit for the specific IT investment. Examples include the following:
    - Configuration or change management control
    - Personnel security
    - Physical security
    - Operations security
    - Privacy training
    - Program/system evaluations whose primary purpose is other than security
    - System administrator functions
    - System upgrades with new features that obviate the need for other stand-alone security controls

- Allocated security control costs for networks that provide some or all necessary security controls for associated applications. Examples include the following:
    - Firewalls
    - IDSs
    - Forensic capabilities
    - Authentication capabilities (*e.g.*, PKI)
    - Additional 'add-on' security considerations.

Ongoing security costs (operations and maintenance costs) are combined with the specific remediation costs and are submitted to OMB in the Exhibit 300s and Exhibit 53 for the budget year. Section 4.7 provides further details on the Exhibit 300 and Exhibit 53.

2.2 Select-Control-Evaluate Investment Life Cycle

In concert with the OMB capital planning and NIST security requirements, agencies are required to adhere to the GAO's best practices, three-phased investment life-cycle model for federal IT investments. As articulated in the GAO's *Information Technology Investment Evaluation Guide*,[11] the three phases—**Select, Control, and Evaluate**—ensure that investment management practices, including security, are disciplined and thorough throughout each phase of the investment life cycle. Figure 2-2 illustrates the three phases.

The **Select** phase refers to activities associated with assessing and prioritizing current and proposed IT projects based on mission needs and improvement priorities and then creating a portfolio of IT projects to address the needs and priorities. Typical Select phase activities include screening new projects; analyzing and ranking all projects based on benefit, cost, and risk criteria; selecting a portfolio of projects; and establishing project review schedules.

---

[11] The guide is available on GAO's Web site at the following address: http://www.gao.gov/policy/itguide/homepage.htm

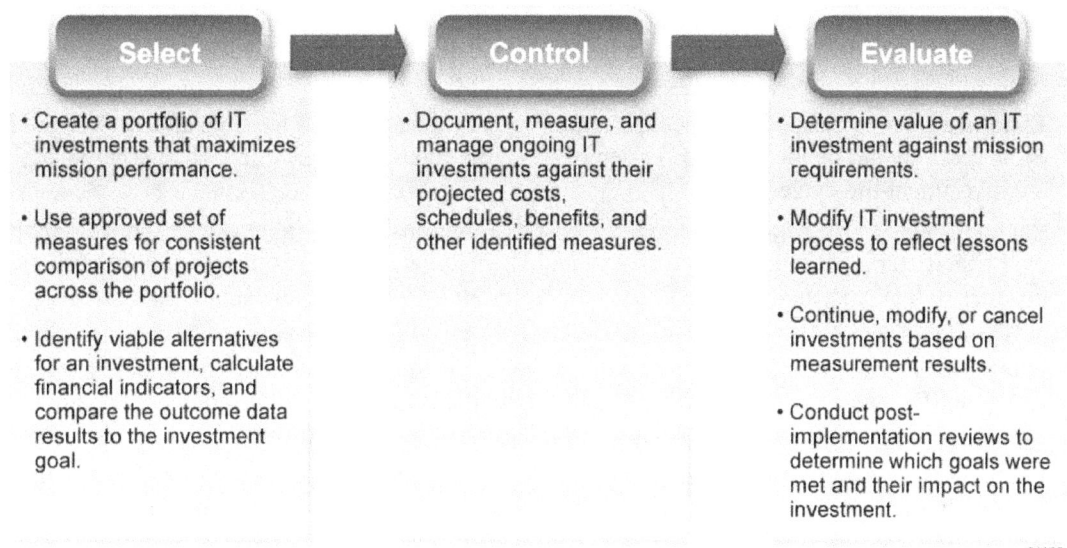

Figure 2-2. The Select-Control-Evaluate Investment Life Cycle

The **Control** phase refers to activities designated to monitor the investment during its operational phase to determine if the investment is within the cost and schedule milestones established at the beginning of the investment life cycle. Typical processes involved in the Control phase include using a set of performance measures to monitor the developmental progress for each IT project to enable early problem identification and resolution.

The **Evaluate** phase refers to determining the efficacy of the investment, answering the question, "Did the investment achieve the desired results and performance goals identified during the Select phase?"

2.2.1 Select Phase

The Select phase typically consists of all of the activities of a business case analysis (BCA). The BCA process is important in the selection of an investment because it enables decision-makers to consider the potential of several investment alternatives before making an acquisition decision. The BCA provides a consistent framework for looking at key variables such as cost of the alternative, benefits the alternative yields, and associated investment risk. These factors can then be compared across a range of alternatives so a single investment alternative can be selected.

A well-prepared BCA incorporates both financial metrics and non-financial factors into a concise and informative presentation. The BCA should also clearly address key issues and facts while revealing the investment's contribution in context to the entire agency and its mission. The following objectives are the components necessary to compose a comprehensive BCA:

- **Evaluate Mission and Objectives**. The BCA should identify the agency's mission and objectives and explain how the investment will enable the agency to fulfill them.
- **Assess Current Environment**. The status quo environment, or the way processes are performed today, should be thoroughly explained in the context of the agency's "to-be" enterprise architecture (EA) blueprint.
- **Perform Gap Analysis**. The BCA should include a discussion of the desired "to be" state. In other words, it should describe the optimal environment to support the agency mission and goals, and point out the necessary steps, procedures, *etc.*, that lie between the status quo and the optimal environment.

- **Identify Investment Alternatives**. The BCA should identify investment alternatives in accordance to budget year Exhibit 300 guidance to reach the optimal environment described in the Gap Analysis.
- **Estimate Cost**. A defined cost element structure should be included for each alternative, and life-cycle costs should be incorporated to demonstrate the financial impact of each alternative across the investment life cycle.
- **Perform Sensitivity Analysis**. Individual cost assumptions and variable values should be adjusted over specified ranges, and the total costs should be estimated. The resulting relationship between changes in total cost and changes in each variable can be quantified to capture the sensitivity of each variable/cost.
- **Characterize Benefits**. Benefits that will accrue as a result of each alternative should be identified and quantified where possible. When quantifiable, the benefits should be compared against life-cycle cost estimates to demonstrate any possible returns on the investment.
- **Perform Risk Analysis**. Investment risk analyses (including security risks) should be conducted for the alternatives, and costs should be adjusted commensurate with anticipated risk.

The objective of a BCA is to measure and illustrate the full impact of an investment within distinct functional areas to make cost and benefit projections on a larger scale. The result of the BCA is clearly the justified selection of a preferred alternative for investment consideration.

2.2.2 Control Phase

Once the preferred alternative is identified and acquisition has taken place, the investment management cycle moves into the Control phase. The primary focus of the Control phase is to document and maintain all current investment processes. During the Control phase, investment owners/project managers should use performance metrics to actively track investment cost and performance at specific milestones as the investment progresses toward meeting its expected mission benefits.

The following management and reporting actions help to improve the accountability and effectiveness of federal investments by establishing, assessing, correcting, and reporting on management controls:

- Review cost metrics and performance indicators
- Assess accountability of results continuously
- Update analyses of each investment's costs and benefits
- Monitor investment costs consistently
- Document all processes and associated costs
- Aggregate cost data and review corrective actions taken to date
- Determine customer satisfaction
- Assess milestone completion/schedule adherence
- Analyze goal achievement
- Evaluate system uptime and other performance indicators as they relate to performance goals.

This disciplined analysis and reporting will allow management to identify and resolve problems early so they can be fixed before they grow into larger problems. For example, management scrutiny and continuous monitoring will facilitate thorough annual FISMA and budget reporting to OMB.

2.2.3 Evaluate Phase

The Evaluate phase assesses the investment's impact and determines future costs for ongoing investments. Annual evaluations determine whether the investment is meeting cost, schedule, and performance goals. At the conclusion of the investment life cycle, the Evaluate phase largely consists of PIR. The feedback and lessons learned generated from the Evaluate phase can be used to refine processes within the Select and Control phases. The following are the key steps to be followed in the Evaluate phase as part of the PIR:

- Compare the investment's actual costs, benefits, risks, and return information against earlier projections and determine the causes of any differences between planned and actual results.
- For each investment in operation, decide whether it should (1) continue operating without adjustment, (2) be further modified to improve performance, (3) be replaced, or (4) be canceled.
- Based on the lessons learned in the PIR, the organization can develop lessons learned and incorporate these findings into its overall investment processes to continue improving the investment management approach.

The data collection and processing steps inherent in the Evaluate phase are an extension of the data recorded in the Control phase. Following the implementation of each investment, reviews should be conducted to determine whether the investments achieved their mission and goals.

2.3 Earned Value Management

Throughout the investment life cycle, agencies should conduct disciplined monitoring to evaluate investment performance and ensure that the investment yields its forecasted benefits to the agency and affected stakeholders. **Earned Value Management (EVM)** is a systematic integration and measurement of cost, schedule, and accomplishments of an investment that enables agencies to evaluate investment performance during **Development, Modernization, and/or Enhancement (D/M/E)**.

In traditional investment management situations, there are two data sources: budgeted (or planned) expenditures and actual expenditures. The comparison of budget versus actual expenditures merely indicates planned spending versus what was actually spent at any given time. This analysis does not address how much has been produced for the amount of money spent or if the investment is maturing according to schedule. Therefore, traditional management approaches do not convey the true cost performance of the investment. However, with EVM, the project manager can identify how much money and time a particular investment is likely to require before selecting it and once selected, how much money was spent at any given time. Furthermore, once selected, the project manager can determine what work has been accomplished to date for the funds expended and how long it will take the investment to reach maturity. Commercial off-the-shelf Earned Value Management Systems (EVMS) or customized tracking systems can be used to monitor earned value metrics across the investment life cycle.

OMB has recognized the utility of EVM for D/M/E investments. OMB Circular A-11[12] requires agencies to use an EVMS along with qualified project managers in order for investment spending to be approved for the fiscal year (FY) and beyond. OMB does not require steady state investments to use EVM.[13] However, OMB Circular A-11 does direct agencies to conduct operational assessments of steady-state investments to demonstrate the investment's performance against cost, schedule, and performance goals. The NIST SP 800-26 self-assessment can be used to perform an operational analysis for IT security.

For annual reporting purposes, agencies are required to document their EVM approach for D/M/E investments in Section I.H., Project (Investment) and Funding Plan, of the annual Exhibit 300. The

---

[12] Based on 2003 guidance
[13] A steady state investment is defined as an asset or part of an asset that has been delivered and is performing the mission.

Exhibit 300 requires project managers to report the investment's progress against the baseline, **Budgeted Cost of Work Scheduled (BCWS), Budgeted Cost of Work Performed (BCWP), and Actual Cost of Work Performed (ACWP)**, as well as the cost, schedule, and performance variance. If project managers do not use EVMS for D/M/E investments, OMB will not give the investment's Exhibit 300 Section I.H. a passing score. Failure to earn a passing score on Section I.H. puts the investment's entire Exhibit 300 at risk for failing and for losing funding. Therefore, EVM is critical not only to investment success but also to securing the funding necessary to acquire and operate IT investments.

While EVM provides a project management approach that provides critical information to the investment management process for enterprise-wide IT security investments, it can also be used to ensure that appropriate security controls are incorporated for D/M/E investments where IT security is embedded into the investment. For example, the work breakdown structure and project plan—required for EVM—should detail when security milestones (*e.g.*, C&A, system security plan completion, security controls testing) will be achieved and what their costs will be. EVM enables the project manager to assess whether the controls are implemented in a timely fashion. The corrective action prioritization framework described in Section 4 identifies how the project manager can use corrective action impacts to determine the level of benefit the security control will provide the investment and the agency. Thus, the combination of EVM and the corrective action prioritization framework can be used to ensure that investment security controls are implemented on schedule and yield the appropriate benefits.

2.4 Information Technology Investment Management

GAO has developed a five-stage model for assessing the maturity of agencies' investment management practices that encompasses portfolio management practices in addition to the Select-Control-Evaluate investment life cycle and the EVM investment evaluation approach. The GAO Information Technology Investment Management (ITIM) maturity framework can be used to determine the current status of an agency's IT investment management capabilities and recommend additional steps an agency can take to strengthen its approach to IT investment management. As agencies work to develop corrective action prioritization techniques, they can use the ITIM framework to help their overall portfolio management techniques to mature.

As referenced in Figure 2-3, each of the five stages builds on the preceding stages and represents increased capabilities and maturing investment management practices.

Each stage is composed of critical processes and key practices that are essential to achieving investment management maturity. IT security is intertwined within the critical practices for each maturity step. As shown in Figure 2-3, as the agency's investment

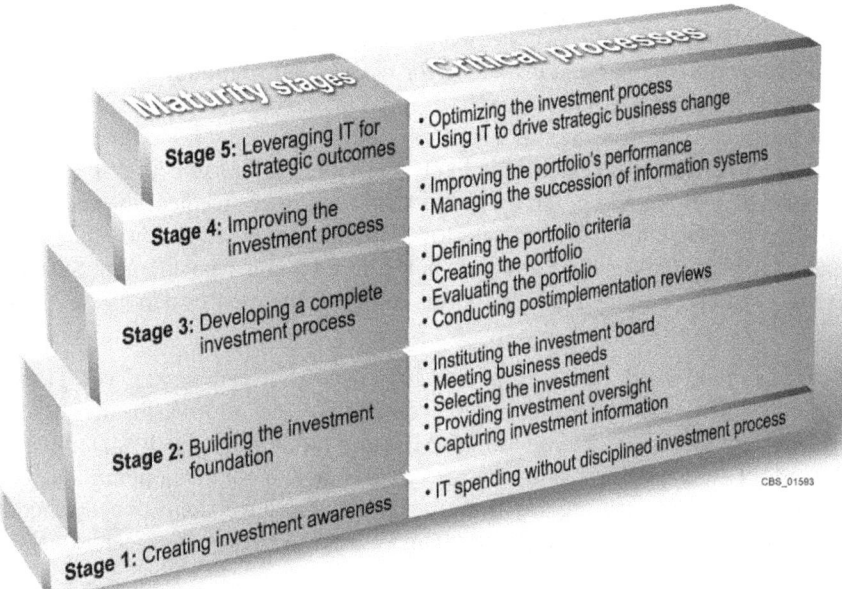

Figure 2-3. ITIM Maturity Model[14]

---

[14] Figure 2-3 is adapted from GAO-04-394-G, *Information Technology Investment Management, A Framework for Assessing and Improving Process Maturity*, Version 1.1, March 2004.

management process matures, its level of project oversight increases. With the increased project oversight formality, investment risk and security become key issues for the individual investments and the entire IT portfolio. Therefore, increased focus on investment and portfolio security and risk levels is required for agencies to advance through the five ITIM stages.

Agencies at Stage 1 maturity level are characterized by unstructured investment management practices. There are no clearly articulated criteria for investment success or failure, leading to unpredictable project outcomes. Agencies at the Stage 1 maturity level are generally assumed to have a basic, inconsistent investment selection process.

As agencies mature to Stage 2 of the ITIM maturity framework, they begin to develop repeatable and sustainable investment selection and control processes. Stage 2 maturity focuses on aligning investments with agency mission and goals; controlling cost, benefit, schedule, and risk (CBSR) milestones to improve project outcomes; and engaging in formal investment reviews from an IRB. Another key aspect of Stage 2 is the creation of a system inventory to ensure the agency can identify CBSR and investment ownership information and review investment performance accordingly.

The system inventory is a cornerstone of the ITIM framework and also relates directly to investment security concerns. Both FISMA and the ITIM framework require the development of a system inventory. FISMA requires the inventory to identify the interfaces between each system and all other systems and networks, including those not operated by or under the control of the agency. The FISMA requirement stems from OMB's expectation that each agency have such an inventory in accordance with its work on developing its EA. FISMA also requires that the inventory be updated at least annually. Agencies should work to build a single system inventory that meets the requirements of both the ITIM framework and FISMA.

Building on the system inventory in Stage 2, Stage 3 moves beyond the Select and Control investment phases and focuses on enterprise portfolio creation, management, and evaluation. The PIR is the principal means for agencies to evaluate their investments' impacts. The PIR is conducted after investment acquisition is completed. The PIR examines the outcome of the investment relative to its plans and expectations. Analyzing PIRs across the investment portfolio enables agencies to identify strong investment management practices and improve their investment management approaches. Also, the increased responsibility of the IRBs builds a foundation for portfolio review and management across the agency. By evaluating IT portfolios across the enterprise, agencies at the Stage 3 maturity level can begin to evaluate the investment's CBSR to determine which portfolio structure will achieve its mission goals while reducing risk to the agency. The information contained in the system inventory provides key indicators for assessing each investment and the portfolio as a whole.

As agencies mature to Stage 4, maturing investment management practices allow agencies to assess whether or not existing investments should continue, be modified, or be canceled. Agencies can also begin to examine new technologies and investments to potentially replace outdated investments and technology.

Agencies that have reached Stage 5, the pinnacle of the ITIM maturity framework, are marked by their ability to learn from other organizations and, more importantly, to continually improve their investment management practices to achieve positive business outcomes. Agencies at Stage 5 are able to benchmark their IT investment processes relative to other "best-in-class" organizations and can proactively evaluate their IT portfolios and emerging investment and technology opportunities to change and improve their overall agency performance.

2.5 Plan of Action and Milestones

Throughout the investment life cycle, the POA&M is used to identify security weaknesses and track mitigation efforts for agency IT investments until the weakness has been successfully mitigated. A robust POA&M process is indicative of increasing ITIM maturity across the agency. OMB requires agencies to prepare and submit POA&Ms for all programs and systems where an IT security weakness has been found. A weakness can be thought of as the gap between current program and system security status and the intended goal/requirement. For example, operating without a contingency plan is a weakness if the system is supposed to have a contingency plan. The POA&M in this example would detail the tasks and milestones to develop, implement, and test a contingency plan.

Prior year (PY) FISMA reporting guidance codifies the exact reporting requirements of the POA&M and should be referenced to ensure the agency is reporting required information to OMB. Table 2-1 contains 10 reporting sections that are typically found in POA&M reporting guidance. However, as previously stated, agencies should reference PY FISMA reporting guidance to ensure they report desired information to OMB. In addition to the POA&M sections listed in Table 2-1, all POA&Ms must contain a unique project identifier. This identifier, which appears in the POA&M, the Exhibit 300, and the Exhibit 53, ties the security costs for the corrective actions in the POA&M to the annual budget information contained in the Exhibit 300 and the Exhibit 53.

Table 2-1. POA&M Sections[15]

POA&M Section	Desired Content
Weakness ID	Provides a unique project identifier or weakness number for each weakness for tracking purposes
Weakness	Refers to a specific identified program or system weakness
Point of Contact (POC)	Identifies the office or organization within the agency that is accountable for correcting the weakness
Resources Required	Details the funding and personnel necessary to mitigate the weakness
Scheduled Completion Date	Indicates corrective action completion date
Milestones with Completion Dates	Refers to major steps that occur while mitigating the corrective action. Timelines and dates are required for each step
Changes to Milestones	Indicates any changes to timelines
Source	Identifies where/how the weakness was identified (e.g., risk assessment)
Status	Indicates if a corrective action is ongoing, delayed, or completed
Comments	Provides space for additional detail or clarification (e.g., causes for delays or potential factors that will impact weakness mitigation)
Security Compliance Gap Percentage	Indicates the difference between the desired and actual compliance with IT security controls. Smaller compliance gap percentages indicate increased levels of IT security compliance

OMB directs agency CIOs and agency program officials to develop, implement, and manage POA&Ms for all programs and systems they operate and control as a part of FISMA compliance. In addition, POA&Ms must be shared with the agency IG to ensure independent evaluation and verification of identified weaknesses and proposed mitigation strategies.

POA&Ms are used at the program level to identify and track weaknesses across enterprise-level initiatives and at the system level to identify and track system-specific weaknesses. Agencies are required to submit POA&Ms to OMB upon request to provide an update on progress against planned remediation efforts. OMB also requires quarterly POA&M update reports to follow the format presented in Figure 2-4. Agencies should reference the latest POA&M reporting guidance from OMB for the most current quarterly reporting requirements and due dates.

---

[15] Note – OMB does not require the "Security Compliance Gap Percentage" entry on POA&M submissions. However, it is useful for agencies to determine the investment's compliance to facilitate prioritization of corrective actions. See Section 4 for an explanation on how to calculate the security compliance gap percentage.

Because the POA&M can be used to track weaknesses at both the agency and system/program level, and it contains the costs/resources necessary to mitigate the identified weaknesses, it is valuable to the corrective action prioritization methodology presented in this guidance. Using the POA&M substantially limits additional data collection because the POA&M contains many of the data points necessary for successful prioritization as discussed in Section 4.

Quarterly POA&M Updated Information		a. Total number of weaknesses identified at the start of the quarter.	b. Number of weaknesses for which corrective action was completed on time by the end of the quarter.	c. Number of weaknesses for which corrective action is ongoing and is on track to complete as originally scheduled.	d. Number of weaknesses for which corrective action has been delayed including a brief explanation for the delay.	e. Number of new weaknesses discovered following the last POA&M update and a brief description of how they were identified (e.g., agency review, IG evaluation, etc.).
Bureau						
	Program Level					
	System Level					
Bureau						
	Program Level					
	System Level					
Total						
	Program Level					
	System Level					

Figure 2-4. POA&M Quarterly Update Format

2.6 Risk

Throughout the investment life cycle, risk management is essential to ensure the investment yields the intended results. While this guidance focuses on security risks, they are only one type of risk that investments face. Therefore, risk management for IT investments can be thought of as a two-pronged approach that includes:

- **Security risks**: risks associated with exploiting IT vulnerabilities and weaknesses
- **Investment risks**: risks associated with the potential inability to achieve overall program objectives within defined cost, schedule, and technical constraints.

In practical implementation, monitoring and mitigating IT security risks is usually at the forefront of IT strategic planning. NIST SP 800-30 contains a comprehensive approach to risk management and can be used as a reference for instituting and maintaining an IT security risk management process. However, monitoring and mitigating investment risks are equally important to successfully achieving intended investment outcomes. Without a strong and consistently applied risk management process, project managers are more likely to:

- Assign inadequate resources to mitigate or resolve major risks
- Make key decisions without adequate information

- Have little insight into potential problems
- Repeat mistakes that plagued earlier projects
- Devote resources to addressing problems rather than avoiding them in the first place
- Fail to deliver a compliant product or service on time and within budget.

Comprehensive risk assessments effectively apply a risk management process that integrates the skills, knowledge, and experience of a variety of specialists to address IT security risks and investment risks. OMB has identified 19 categories of risk. They are:[16]

- **Schedule**: Risk associated with schedule slippages, either from lack of internal controls or from those associated with late delivery by vendors, resulting in missed milestones.
- **Initial costs**: Risk associated with "cost creep" or miscalculation of initial costs that result in an inaccurate baseline against which to estimate and compare future costs.
- **Life-cycle costs**: Risk associated with misestimating life-cycle costs and exceeding forecasts and relying on a small number of vendors without sufficient cost controls.
- **Technical obsolescence**: Risk associated with technology that becomes obsolete before the completion of the life cycle and cannot provide the planned and desired functionality.
- **Feasibility**: Risk that the proposed alternative fails to result in the desired technological outcomes; risk that business goals of the program or initiative will not be achieved; risk that the program effectiveness targeted by the project will not be achieved.
- **Reliability of systems**: Risk associated with vulnerability/integrity of systems.
- **Dependencies and interoperability between this investment and others**: Risk associated with interoperability between other investments; risk that interoperable systems will not achieve desired outcomes; risk of increased vulnerabilities among systems.
- **Surety (asset protection) considerations**: Risk associated with the loss/misuse of data or information; risk of technical problems/failures with applications; risk associated with the security/vulnerability of systems.
- **Risk of creating a monopoly for future procurements**: Risk associated with choosing an investment that depends on other technologies or applications that require future procurements to be from a particular vendor or supplier.
- **Capability of agency to manage the investment**: Risk of financial management of investment, poor operational and technical controls, or reliance on vendors without appropriate cost, technical, and operational controls; risk that business goals of the program or initiative will not be achieved; risk that the program effectiveness targeted by the project will not be achieved.
- **Overall risk of project failure**: Risk that the project/investment will not result in the desired outcomes.
- **Project resources/financial**: Risk associated with "cost creep," miscalculation of life-cycle costs, reliance on a small number of vendors without cost controls, or inadequate acquisition planning.
- **Technical/technology**: Risk associated with immaturity of commercially available technology and reliance on a small number of vendors; risk of technical problems/failures with applications and their inability to provide planned and desired technical functionality.
- **Business/operational**: Risk associated with business goals; risk that the proposed alternative fails to result in process efficiencies and streamlining; risk that business goals of the program or initiative will not be achieved; risk that the investment will not achieve operational goals; risk that the program effectiveness targeted by the project will not be achieved.

---

[16] Risk definitions are a combination of OMB-provided definitions and best practices definitions.

- **Organizational and change management**: Risk associated with organizational-, agency-, or government-wide cultural resistance to change and standardization; risk associated with bypassing, lack/improper use of, or non-adherence to new systems and processes because of organizational structure and culture; risk associated with inadequate training planning.

- **Data/information**: Risk associated with the loss or misuse of data or information; risk of compromise of citizen or corporate privacy information; risk of increased burdens on citizens and businesses because of data collection requirements if the associated business processes or project (being described in the Exhibit 300) requires access to data from other sources (federal, state, and/or local agencies).

- **Security**: Risk associated with the security/vulnerability of systems, Web sites, and information and networks; risk of intrusions and connectivity to other (vulnerable) systems; risk associated with the evolution of credible threats; risk associated with the criminal/fraudulent misuse of information; must include level of risk (high, moderate, low) and what aspect of security determines the level of risk (*e.g.*, need for confidentiality of information associated with the project/system, availability of the information or system, or integrity of the information or system).

- **Strategic**: Risk associated with strategic- and government-wide goals (*e.g.*, President's Management Agenda [PMA] and e-Gov initiative goals); risk that the proposed alternative fails to result in achieving those goals or in making contributions to them.

- **Privacy**: Risk associated with the vulnerability of information collected on individuals or risk of vulnerability of proprietary information on businesses.

In addition to the 19 OMB-defined areas of risk, a comprehensive risk assessment should include the assessment of each of the following categories:

- **Product Risk Assessment**: Identifies those risks associated with a given system concept. This technique is used to identify and analyze risks in the following critical risk areas: design and engineering, technology, logistics, production, concurrency, and both hardware and software.

- **Process Risk Assessment**: Analyzes program technical risks resulting from the contractor's processes. The primary benefit of this assessment addresses pervasive and important sources of risk in most investments.

- **Threat and Requirements Risk Assessment**: Assesses risks related to risk drivers. To a large degree, operational needs, environmental demands, and threats determine system performance requirements. They are a major factor in driving the design of the system and can introduce risk in an investment.

- **Cost Risk Assessment**: Provides an investment-level cost-estimate-at-completion that is a function of performance and schedule risks.

- **Quantified Schedule Risk Assessment**: Provides a means to determine investment-level schedule risk as a function of risk associated with various activities that compose the investment.

All risks contribute to the calculation of **risk-adjusted cost**, which OMB now requires agencies to report in each investment's Exhibit 300. The risk-adjusted cost calculation provides a range of how the investment's costs will be affected if part or all the investment and security risks identified in the Exhibit 300 manifest themselves. The risk-adjusted costs provide realistic forecasts across the investment life cycle, allowing decision-makers to plan appropriately for risks to the investment. The forecasts will also allow OMB to determine whether the risk-adjusted cost precludes the investment from receiving funding because of the potential financial burden caused by investment risks.

3. IT Security and Capital Planning Integration Roles and Responsibilities

Integrating IT security into the capital planning process requires input and collaboration across agencies and functions. Figure 3-1 depicts a hierarchical approach to capital planning in which investment decisions are made at both the enterprise and operating unit levels. A certain level of investment management maturity is assumed in the framework presented in Figure 3-1

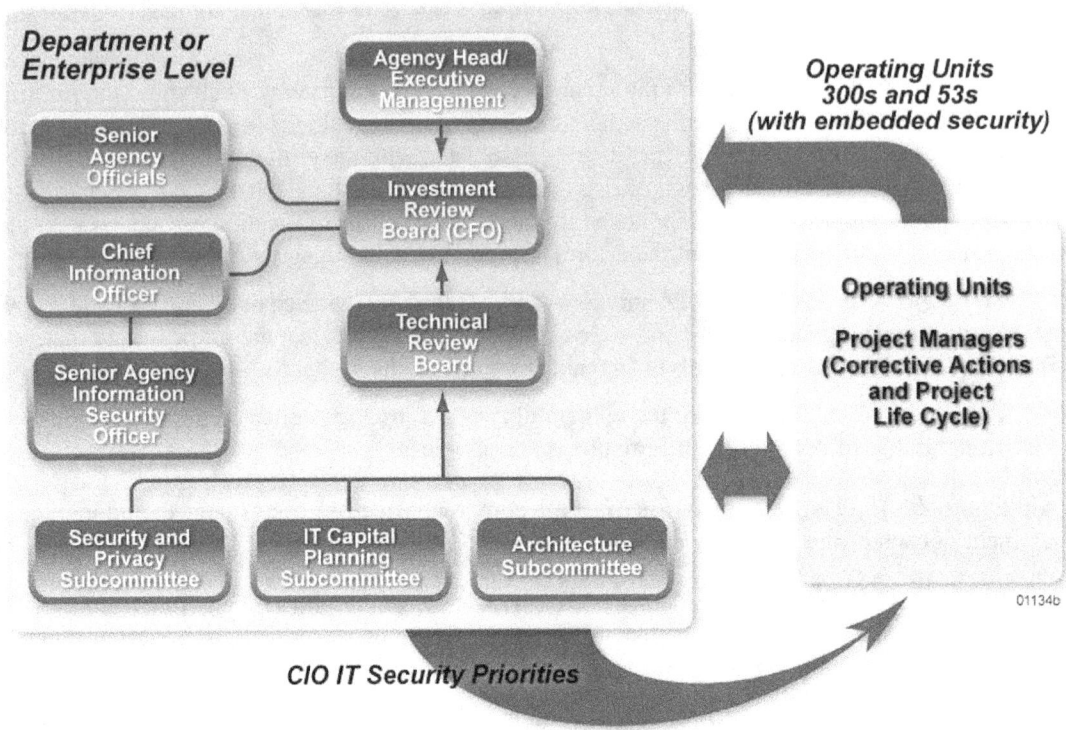

Figure 3-1. Notional IT Management Hierarchy

While specific practices for investment management vary greatly at the operating unit level because of varying sizes and missions of the operating units, the process generally mirrors the process at the departmental level. The CIO formulates and articulates IT security priorities to the organization to be considered within the context of all agency investments. Priorities may be based on agency mission, executive branch guidance such as the PMA, OMB guidance, or other external/internal priorities. Examples of security priorities include certifying and accrediting all systems or implementing PKI throughout the enterprise. (It is important to note that OMB/Executive Branch guidance or laws should be ranked highest among these priorities.)

Once operating units finalize their IT portfolios and budget requests for the budget year, they forward their requests to the agency-level decision makers. At the agency level, several committees evaluate IT portfolios from the operating units as referenced in Figure 3-1, culminating in a review by the IRB. The IRB then decides on an agency-level IT portfolio and forwards recommendations to the agency head for review. Once the agency-level IT portfolio is approved by the agency head, the necessary Exhibit 300s and Exhibit 53 are forwarded to OMB to obtain funding.

Generally, project managers in operating units manage investments according to federal and agency policies, the CIO-articulated priorities, and specific operating unit priorities. Project managers also identify vulnerabilities and needed corrective actions for their investments. Each year, project managers prepare and submit Exhibit 300s to the operating unit management and operating unit IRBs. These

Exhibit 300s for mixed life-cycle and steady-state investments are combined with Exhibit 300s for new investments and are prioritized at the operating unit level to determine the appropriate IT portfolio mix for the budget year.

The described IT management framework will vary from agency to agency. The important element common to all agencies, though, should be standardized approval hierarchies and parallel planning and prioritization processes at both the enterprise and operating unit levels.

3.1 Overview

Many different stakeholders from IT security, capital planning, and executive leadership areas play roles and make decisions on integrating IT security into the capital planning process and ultimately forming a well-balanced IT portfolio. Figure 3-2 illustrates the roles and responsibilities hierarchy for integrating IT security into the CPIC process. While specific roles and responsibilities will vary from agency to agency, involvement at the enterprise and operating unit levels throughout the process allows agencies to ensure that capital planning and IT security goals and objectives are met. Figure 3-2 identifies leading, supporting, or approving roles for each stakeholder as they apply to the integration of security into the CPIC process phases.

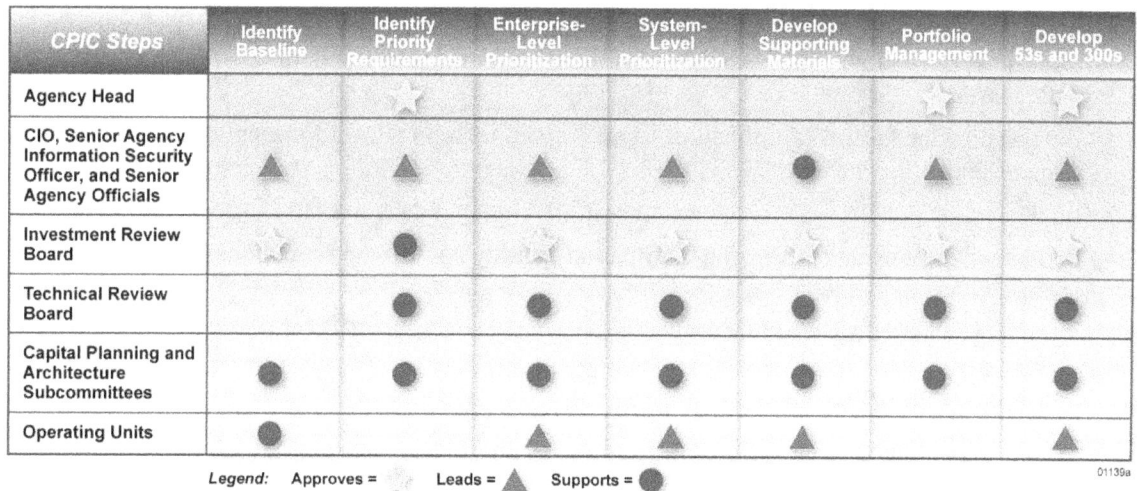

Figure 3-2. Roles and Responsibilities Throughout the CPIC Process

Sections 3.2 through 3.12 describe stakeholder roles and responsibilities.

3.2 Head of Agency

FISMA charges the agency head with ensuring appropriate agency security posture and with reporting to Congress on the status of agency security posture. This position oversees the security policy and the resource budget and has ultimate management responsibility for resource allocation. The agency head has the following responsibilities related to integrating IT security into the CPIC process:

- Complying with FISMA requirements and the related information resource management policies and guidance including OMB Circular A-130 established by the Director of OMB and the related IT standards promulgated by the Secretary of Commerce
- Ensuring that established information security and resource management policies and guidance are integrated with agency strategic and operational planning processes under FISMA and are communicated promptly and effectively to all relevant agency officials
- Ensuring that senior agency officials provide information security for the information and information systems that support the operations and assets under their control

- Establishing strategic agency mission and vision (establishing goals which flow down to budget, IT, and security priorities) and ensuring that information security management processes are seamlessly integrated into those processes and documents
- Ensuring that the information protection is commensurate with risk and magnitude of harm resulting from the information's compromise
- Approving the overall annual IT budgets and overall portfolio (with appropriate security integrated) developed through the IRB process
- Establishing priorities to achieve improvements that comply with the PMA
- Delegating the authority to ensure compliance with agency security requirements to the agency CIO.

3.3 Senior Agency Officials

Under the direction of the agency head, senior agency officials provide information security for the data and IT systems that support the operations and assets under their control. The senior agency official duties include:

- Assessing the risk and magnitude of the harm that could result from the unauthorized access, use, disclosure, disruption, modification, or destruction of information and information systems under their control
- Determining the levels of information security appropriate to protect information and information systems under their control
- Implementing policies and procedures to cost-effectively reduce risks to an acceptable level
- Periodically testing and evaluating information security controls and techniques to ensure that they are effectively implemented
- Providing senior IT advice to the head of each agency and to the IRB.

3.4 Chief Information Officer

The Information Technology Management Reform Act of 1996 (also known as the Clinger-Cohen Act) requires agencies to appoint CIOs. The agency CIO is the senior IT advisor to the IRB and the head of the agency. In this capacity, the CIO role includes:

- Assisting senior agency officials with IT issues
- Developing and maintaining an agency-wide information security program
- Developing and maintaining risk-based information security policies, procedures, and control techniques
- Designating a senior agency information security officer (ISO) to carry out CIO directives as required by FISMA, including POA&M responsibilities
- Designing, implementing, and maintaining processes for maximizing the value and managing the risks of IT acquisitions
- Presenting proposed IT portfolios to the IRB
- Providing final portfolio endorsements
- Presenting and recommending Control and Evaluate decisions and recommendations.

3.5 Senior Agency Information Security Officer

As mandated by FISMA, the senior agency ISO is appointed by the CIO and manages information security throughout the agency. The senior agency ISO is responsible for coordinating program requirements throughout the agency with designated POCs and project managers. Their duties include:

- Developing and maintaining an agency-wide information security program
- Issuing annual IT security planning guidance, including security priorities, objectives, and prioritization criteria for new and legacy systems
- Training and overseeing personnel with significant responsibilities for information security with respect to such responsibilities
- Developing and maintaining information security policies, procedures, and control techniques
- Assisting senior agency officials concerning their IT security-related responsibilities.

3.6 Chief Financial Officer

As a member of the IRB, the agency CFO is the senior financial advisor to the IRB and the head of the agency. In this capacity, the CFO is responsible for:

- Reviewing the cost goals of each major investment
- Reporting financial management information to OMB as part of the President's budget
- Complying with legislative and OMB-defined responsibilities as they relate to IT capital investments
- Reviewing systems that impact financial management activities
- Forwarding personal investment assessments for review by the entire IRB.

3.7 Investment Review Board

Composed of the CFO and senior managers at the agency or operating unit level, the members of the IRB evaluate existing and proposed IT investments to determine the appropriate mix of investments that will allow the agency to achieve its goals. In this capacity, IRB duties include:

- Operating at the enterprise level
- Approving the CIO's IT strategic guidance, including security priorities and prioritization criteria; these priorities and criteria need to reflect the evolving security needs
- Approving corrective action prioritization
- Approving the controls and evaluating the IT portfolio with embedded security requirements, objectives, measures, and milestones
- Ensuring alignment of the PMA achievement and strategic agency missions and vision with IT security priorities and criteria.

3.8 Technical Review Board

The Technical Review Board (TRB) is composed of IT security and architecture managers from the Office of the CIO (OCIO) and other applicable members. The TRB's duties include:

- Conducting detailed IT investment review and security analyses and reviewing business cases for security requirements
- Balancing IT investment portfolios based on CIO/IRB IT security priorities and prioritization criteria

- Acting as a focal point for agency coordination of OCIO strategic planning, architectural standards, and outreach to organizations and bureaus.

3.9 IT Capital Planning, Architecture, and Security and Privacy Subcommittees

The subcommittees provide subject matter expertise and advice to the OCIO and operating units. In this capacity, the subcommittees are responsible for:

- Translating OMB IT capital planning security guidance into operational and internal process control enhancements
- Supplying process improvements and providing EA support for the TRB.

3.10 Operating Unit/Bureau Executive Management

As representatives of their respective operating units/bureaus within the IRB, operating unit/bureau executive management focuses on the process for integrating IT security priorities into business cases and the OMB Exhibit 53/300 process.

3.11 Project Manager

The project manager has overall responsibility for coordinating the management and technical aspects of a system's life cycle. Project manager responsibilities include the following:

- Developing a project management plan that integrates security throughout the life cycle
- Developing a cost and schedule baseline and completing a project within schedule and budget constraints while meeting the customer's needs
- Coordinating the development, implementation, and operation and maintenance of a system with appropriate units within an agency
- Reporting the results of projects to the system owner and other appropriate agency staff
- Presenting, when appropriate, the progress of critical projects to the OCIO, the IRB, and other applicable review entities.

3.12 System Owner

The system owner handles the day-to-day management of the IT investment. The system owner responsibilities include the following:

- Maintaining active senior-level involvement throughout the development of the system
- Participating in project review activities and reviewing project deliverables
- Coordinating activities with senior management
- Obtaining and managing the budget throughout the project's life cycle against a project manager's delivered, locked baseline
- Holding review and approval authority to ensure that developed products incorporate security and meet user requirements
- Ensuring system has an up-to-date security plan, has a contingency plan, and receives full C&A
- Providing baseline assessment performance measures to evaluate the security of the delivered IT initiative
- Developing and maintaining system-specific POA&Ms.

4. Integration of Security into the CPIC Process

The CPIC process is defined by OMB Circular A-130 as "a management process for ongoing identification, selection, control, and evaluation of investments in information resources. The process links budget formulation and execution, and is focused on agency missions and achieving specific program outcomes." Integrating security into this process ensures that information resources are planned and provided for in a thorough, disciplined manner, ultimately enabling improved security for IT investments.

4.1 The CPIC Process and IT Security

Integrating security into the CPIC process consists of a seven-step methodology to ensure that mission and security requirements are met throughout the investment life cycle. Figure 4-1 depicts the seven-step methodology of integrating security into the CPIC process

As Figure 4-1 indicates, key activities and decisions take place throughout the CPIC process to ensure that security requirements are identified, planned for, and implemented as a part of an individual IT investment or the overall agency investment portfolio. The first step is to identify the security baseline using IT security metrics to determine where security weaknesses exist. Following identification of the baseline, prioritization requirements are established. Corrective actions to mitigate vulnerabilities must be evaluated against the security requirements. Requirements can be CIO-articulated security priorities, enterprise-wide initiatives, or as shown in the example in this guidance document, NIST SP 800-26 topic areas. Following identification of the prioritization criteria, corrective actions should be prioritized against the criteria on the basis of cost and impact, first at the enterprise level and then at the system level. Once corrective actions have been prioritized, business cases and Exhibit 300s should be developed and submitted to the IRB for inclusion in the agency IT investment portfolio. The IRB then prioritizes business cases at the agency level and determines the agency IT investment portfolio and funding levels necessary for submission to OMB via Exhibit 53 and the Exhibit 300s. Once funding allocations are determined, managers must manage their investments to the cost, schedule, and performance goals set forth in the Exhibit 300.

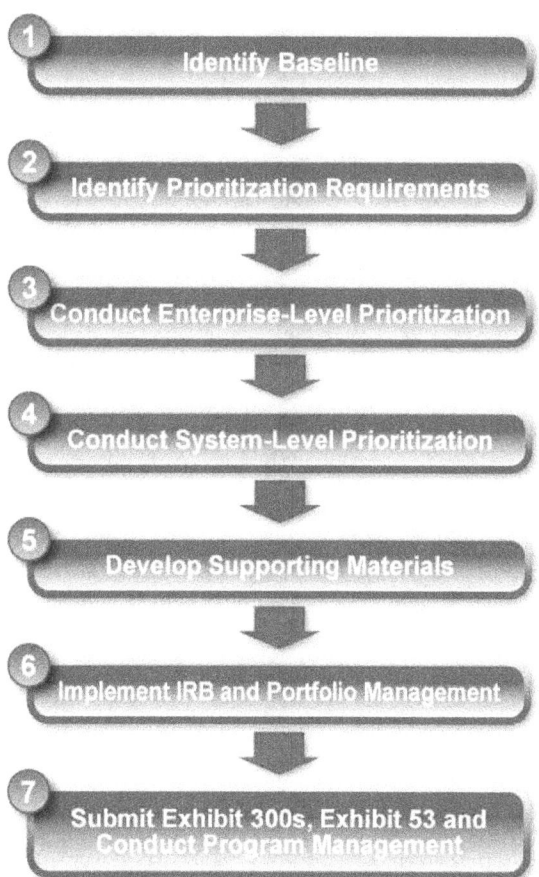

Figure 4-1. Integrating IT Security Into the CPIC Process

The activities and decisions that occur throughout the CPIC process relate to critical milestones in the investment life cycle. Figure 4-2 maps the CPIC steps and activities to each phase of the Select-Control-Evaluate investment life cycle.

	Select	Control	Evaluate
Identify Baseline	Agency-specific and federally mandated security requirements are identified. Metrics are used to determine agency's compliance with requirements.	Project managers monitor investment performance against baseline to ensure no new corrective actions are necessary to mitigate security compliance gaps.	Post Implementation Reviews are conducted to determine whether the investment has achieved its desired results.
Identify Prioritization Requirements	Agency CIO or other senior management official articulates IT security themes based on agency mission goals and priorities.	Agency CIO or other senior management official articulates IT security themes based on agency mission goals and priorities.	
Prioritize Against Themes	Project managers prioritize investments based on CIO - articulated themes and score initiatives against each other on specific risk, financial, technological, management, legislative, and OMB requirements.	Project managers monitor initiatives against the changing CIO's IT security themes and performance criteria to determine whether they are achieving anticipated results and if any modifications are necessary.	Metrics are used to evaluate investment performance against the baseline.
Develop Supporting Materials	Project managers develop business cases and other supporting materials to support initiatives aligned to agency mission and goals.	Project managers ensure their initiatives remain aligned with evolving mission and goals as the initiatives mature.	
Perform Portfolio Management - ITIRB	The ITIRB reviews completed business cases and selects an appropriate mix of investments for inclusion in the agency's portfolio. ITIRB can use prioritized corrective actions as a selection criterion. Successfully scored initiatives are added to the OMB investment pool, pending funding approval from OMB.	The ITIRB reassesses initiatives that have undergone major modifications and monitors costs and security performance.	
Develop Exhibit 53s, 300s, Program Management	Project managers prepare and submit Exhibit 300 and Exhibit 53 budget request/justification materials and submit the documentation to OMB for review and funding approval.	Project managers review and update Exhibit 300s annually to reflect changes in the investment.	

Figure 4-2. IT Security Activities Throughout the Investment Life Cycle

Figure 4-3 illustrates the relationship among the security drivers, the investment life cycle, and the System Development Life Cycle (SDLC).

During the **Select** phase, security drivers include assessment activities to ensure that IT security investments comply with security requirements, discussed in Section 2.1. During the **Control** phase, investments are monitored through the use of security metrics to ensure that security controls are in place and operational and that investments remain compliant with requirements. During the **Evaluate** phase, security drivers include self-assessment activities to ensure compliance and media sanitization efforts following removal from operation and prior to disposition.

Investment Lifecycle	Select		Control	Evaluate	
SDLC Phases	Initiation	Acquisition/ Development	Implementation	Operations/ Maintenance	Disposition
Security Considerations	• Data Sensitivity Analysis • Privacy Impact Assessment	• Risk Assessment • System Security Plan • POA&M • Security Controls • Contingency Planning • Security Test and Evaluation • Certification	• Approval/ Authorization to Operate	• Continuous Monitoring Using Metrics • Periodic Reviews • Annual Self-Assessment • Recertification	• Data Sensitivity Analysis • Privacy Impact Assessment

Figure 4-3. Investment Life Cycle and SDLC Decision-Making

4.2 Identify Baseline

The first step of the CPIC process is to assess the security baseline. The security baseline provides a snapshot of the agency's compliance with baseline security requirements (BLSR) and is instrumental in identifying IT security strengths and weaknesses. The result of a security baseline analysis enables agency executives to evaluate their IT security posture and identify areas for improvement.

Agencies can identify their baseline at two levels:

1. The investment-level baseline evaluates each IT investment's compliance with regulations and the overall security posture at the system level.

2. The enterprise-level baseline aggregates the investment-level results to provide an overall security posture assessment across the agency.

As Figure 4-4 illustrates, an information security metrics program is the best way to define the security baseline. NIST SP 800-55, *Security Metrics Guide for Information Technology Systems*, provides guidance on developing and implementing an information security metrics program. Information security metrics programs use existing data sources to create a quantifiable picture of the security posture of individual IT investments and then aggregate that data to provide an overview across the organization. Metrics programs can assess IT security investments' compliance with NIST SP 800-26 topic areas, FISMA requirements, and agency regulations. Metrics can provide compliance percentages that indicate the existence of adequate security controls, highlight current weaknesses, and identify gaps between actual and desired IT security controls implementation. Information about these gaps provides inputs into requirements for mitigation efforts, such as identifying corrective actions that mitigate the vulnerabilities and improve the security controls across the agency.

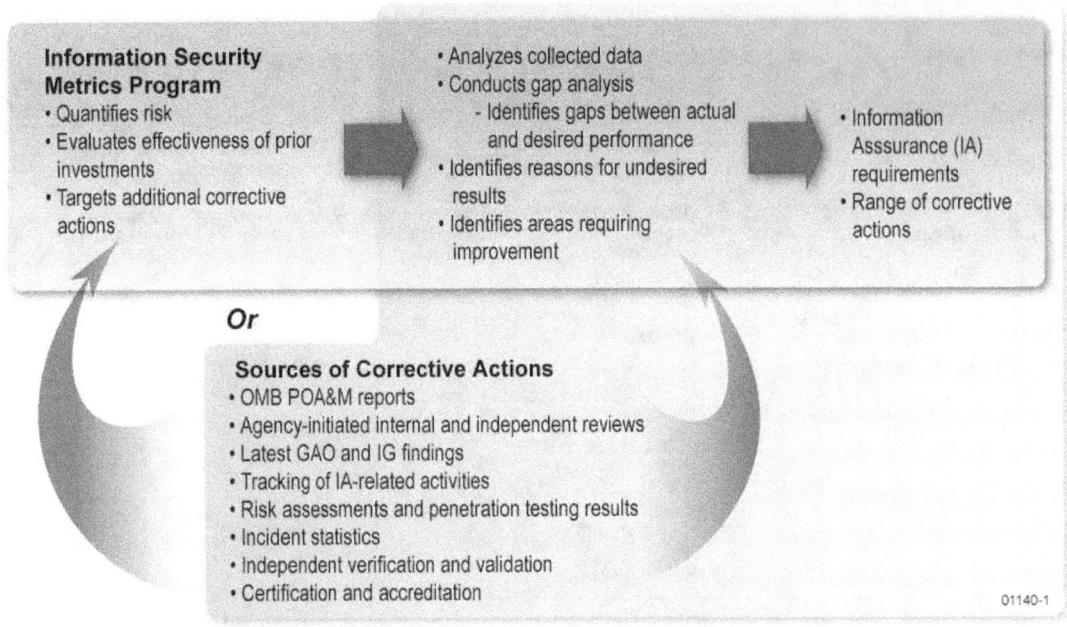

Figure 4-4. Identifying Baseline Best Practices

If an agency does not have a mature or robust IT security metrics program, there are other sources of information that can be used to establish the IT security baseline. OMB POA&Ms, agency-specific security reviews, GAO and IG audit findings, risk assessments, incident statistics, and other information can be assembled to provide an overview of an agency's security posture showing where security investments are needed. The advantages of a mature metrics program are the simplification of data collection, reporting versus consolidating information from a variety of assessments, and the ability to trend results over time. A mature metrics program is also more cost effective than ad hoc data collection from multiple data sources. Whether the agency uses an IT security metrics program or other methods for assessing the information security baseline, the output remains consistent: an understanding of the strengths, weaknesses, and vulnerabilities that exist within an agency's security controls that underscore where investments are required. The resulting vulnerabilities and weaknesses then serve as inputs into the next step of the CPIC process: identifying prioritization criteria.

4.3 Identify Prioritization Criteria

To identify appropriate prioritization criteria, agencies should use the security baseline in determining how to allocate resources. Each agency should implement corrective actions for each identified vulnerability and weakness to ensure its IT portfolio complies with federal mandates and demonstrates security controls commensurate with the sensitivity of each investment. However, available funding does not always allow all security baseline needs assessment requirements to be addressed immediately. Therefore, requirements must be prioritized to address the most pressing security investment needs first. Without a systematic risk management approach, such prioritization can be difficult.

Corrective actions should be prioritized to ensure the most efficient use of resources, including personnel time and mitigation costs. However, to effectively prioritize corrective actions, agencies must identify criteria for prioritization. Specific prioritization criteria will vary from agency to agency; however, the common approach is to rank order IT security investments.

Examples of general prioritization criteria include:
- Federal government priorities
    - PMA
    - FEA requirements
    - e-Government scorecard
    - Compliance with rules and regulations—Clinger-Cohen Act, Homeland Security Presidential Directives, FISMA, and Health Insurance Portability and Accountability Act
    - NIST standards and guidance
    - OMB requirements
- Agency mission and goals that align with specific agency concerns and its risk profile
- Government and agency initiatives include, for example:
    - E-authentication
    - E-tax filing
    - E-clearance
    - E-grants.

> Agencies will undoubtedly discover vulnerabilities from the baseline security assessment that require immediate attention and, therefore, should be subject to immediate action. For example, if an agency discovered that perimeter firewalls were improperly configured, allowing unauthorized external access to sensitive information, the agency should immediately allocate resources to mitigate the vulnerability before engaging in the prioritization exercise. Therefore, following the baseline assessment, the agency should immediately allocate resources (both personnel and financial) to mitigate pressing vulnerabilities that put the agency at immediate risk. The remaining findings should then be prioritized to make efficient use of remaining resources.

Other criteria include IT security priorities. Security priorities embody an agency's approach to IT security. They are articulated and promulgated by the agency CIO or other senior management officials, and evolve over time to reflect the changing maturity level of the agency's security program. Security priorities can be proactive in an effort to stay ahead of potential threats, or they can be reactive to comply with legislative requirements. Examples of IT security priorities include:

- Complying with statutory requirements in Clinger-Cohen Act, FISMA, and OMB A-130 guidance
- Implementing a risk-based security program (FISMA and Executive Orders)
- Safeguarding national and agency mission-critical assets (Homeland Security Presidential Directives)
- Improving Information Security Program status
- Completing C&A of all systems in accordance with NIST guidance and standards.[17]

Agency operating units will develop their IT security investment strategies in alignment with the CIO-articulated IT security priorities and agency-identified prioritization criteria. When combined, prioritization criteria and CIO-articulated priorities form specific prioritization frameworks that allow investments to be rank-ordered against requirements. Examples of specific prioritization frameworks include:

- Strategic view
    - Linkage with a government-wide initiative
    - Impact on agency goals
    - Impact on e-Government scorecard improvement
    - Mission criticality

- IT security view
    - Support of agency mission: system and information impact
    - FIPS 199
    - Security controls:
        - ✓ NIST SP 800-26 topic areas or critical elements
        - ✓ NIST SP 800-53 control families
        - ✓ Similar agency-specific framework

---

[17] See NIST SP 800-37, *Guide for the Security Certification and Accreditation of Federal Information Technology Systems*.

- Results
  - ✓ Improvement in compliance with regulations
  - ✓ Reduced cost of implementation
  - ✓ Acceptance of residual risk
- Impact on security posture
  - ✓ Magnitude of impact on the overall agency security posture
  - ✓ Cost-effectiveness of the action ("bang for the buck").

While any of these criteria could be used to prioritize corrective actions, this guidance document uses the 17 NIST SP 800-26 topic areas for that purpose. The 17 topic areas articulate a common body of security that should be present for any federal IT system. The POA&M communicates security weaknesses that are discovered during the self-assessment, security reviews, audits, and other similar activities and the corresponding corrective actions. These corrective actions can be easily categorized into the 17 topic areas to provide a basis for prioritization. Therefore, because they reuse existing data and provide a common body of security controls, the 17 NIST SP 800-26 topic areas and the POA&M data are ideal mechanisms for ranking and prioritizing corrective actions.

> The use of **NIST SP 800-26** topic areas as prioritization criteria in this guidance is purely for example purposes. As referenced above, there are many other prioritization criteria that can be used to prioritize corrective actions. The security control families in **NIST SP 800-53** provide another example of possible prioritization criteria.
>
> Each agency can make the determination of which criteria will work best given their unique operating environments. No matter which prioritization criteria an agency chooses, it can be substituted into the methodology presented in this guidance.

4.4 Prioritize Against Requirements

Once agency management and stakeholders agree on prioritization requirements, the agency must begin the prioritizing process by rank-ordering requirements against the prioritization criteria. The objective is to apply the first security dollar to the most critical security investment. The next dollar is then applied to the next critical security investment and so forth until the security budget is expended.

To determine a rank order of topic areas for an agency, multiple stakeholders should meet and agree on the prioritization scheme. Stakeholders may include the CIO, senior security officials, key system owners, and members of the IRB. A high degree of coordination is required to perform this activity successfully to ensure buy-in from all parties. It is important to bring stakeholders together early in the process and involve them throughout the process. It also may be helpful to have a facilitated session using a decision support tool to coordinate input from multiple parties.

After determining the system- and enterprise-level baselines and after establishing prioritization criteria, an agency can prioritize corrective actions at the same two levels:

1. **System-level prioritization** – prioritize corrective actions to address system-level security weaknesses and vulnerabilities found during the baseline assessment against the predefined prioritization criteria. This prioritization occurs at the operating unit level by system owners and project managers. This information should be available from an agency's POA&M.

2. **Enterprise-level prioritization** – prioritize enterprise-wide security corrective actions identified during the baseline assessment based on predefined prioritization criteria. This prioritization occurs at the enterprise level by senior agency officials.

In addition to prioritizing requirements at the system and enterprise levels, two other data points are needed for the prioritization process:

1. **Compliance gap** – the difference between the desired and actual compliance with the security requirements. For example, if an IT system has completed 80 percent of C&A activities, that investment would have a C&A compliance gap of 20 percent. (The actual compliance of 80 percent is subtracted from the desired compliance of 100 percent to yield a 20 percent compliance

gap.) The smaller the compliance gap, the more compliant the system or enterprise control. This information is part of the FISMA report.

2. **Corrective action impact** – the ratio of compliance gap to corrective action cost. The formula for corrective action impact appears in Figure 4-5. As illustrated, the corrective action impact is calculated by dividing the compliance gap percentage by the cost to implement the corresponding corrective action(s). This ratio provides a proportion of result to cost. The higher the impact proportion, the more "bang for the buck" the corrective action will provide. The resulting proportion is multiplied by 100,000 to facilitate further calculations.

$$\left(\frac{Corrective\ Action\ Security\ Compliance\ Gap\ \%}{Corrective\ Action\ Cost}\right) \times 100,000$$

Figure 4-5. Corrective Action Impact

As an example of the corrective action impact, if an IT investment has completed 80 percent of C&A activities, and the remaining cost to complete C&A is $150,000, the corrective action impact would be:

$$\left(\frac{20\ \%}{\$150,000}\right) \times 100,000 = .13$$

NIST SP 800-26 assesses the current security posture of information systems. As a result, it is a good benchmark for assessing system and agency security and is used in the example in this guidance. If an agency determined that one of its prioritization requirements was to comply with the 17 NIST SP 800-26 topic areas, and another prioritization criteria was to address the most sensitive systems first, it would need to:

- Rank-order the topic areas in order of importance to the agency
- Rank-order agency systems according to FIPS 199 category
- Calculate the compliance gaps at the enterprise and investment levels
- Calculate the corrective action impact at the enterprise and investment levels.

These four steps are articulated in greater detail in the subsequent sections of this guidance.

4.4.1 Enterprise-level Prioritization

To repeat the example above, if an agency determined that its prioritization requirements were to be compliant with the 17 NIST SP 800-26 topic areas and to address the most sensitive systems first, it would need to conduct prioritization at the enterprise level first to determine which NIST SP 800-26 topic areas should be addressed across the agency.

To begin the process, agency executives should have already rank-ordered the 17 NIST SP 800-26 topic areas using an analytical hierarchy tool. For the purposes of this exercise, agency executives should rank the topic areas into one of three categories—high, moderate, and low—based on the topic area's importance to agency mission and goals, and in the context of any CIO-articulated priorities and enterprise-level initiatives. While every topic area should be of some importance to the agency, in a resource-constrained environment, each area should be prioritized according to the overall agency priorities into high, moderate, and low groups. There are no exacting standards for delineation; agencies should work within their own environment to determine the appropriate number of high, moderate, and low topic areas.

Following the ranking of the 17 topic areas, agencies should determine the aggregate compliance for each topic area. This information already exists in each system's self-assessment, so this exercise is simply reusing existing data. Once compliance percentages are obtained for each system, the results should be aggregated across the agency for an overall compliance percentage. Then, the security compliance gap percentage is calculated by subtracting (from 100 percent) the average compliance across the agency for each topic area.

The next step is to determine the cost to implement the corrective action. This information is found in the POA&M; therefore, no new information is needed. For the enterprise-level prioritization, the dollar amounts for each topic area should be aggregated to obtain a total across the agency. In the event that this information is not included in the POA&M, it will need to be calculated. There are several ways to derive costs for POA&M corrective actions. Costing software tools can perform this cost-estimating function, or the agency can rely on historical prices or existing relationships with vendors to determine corrective action costs.

Once the security compliance percentage gap and the corrective action costs are obtained, the corrective action impact can be calculated. As referenced earlier, the corrective action impact is calculated as follows:

$$\left(\frac{Corrective\ Action\ Security\ Compliance\ Gap\ \%}{Corrective\ Action\ Cost} \right) \times 100,000$$

Finally, after calculating the corrective action impact, agency executives should prioritize the corrective actions for each topic area according to the corrective action impact. The impact should be delineated into three categories: great, average, and basic. Agency stakeholders spanning the roles of line operations to agency executives should determine the boundaries for each of the three categories. For example, stakeholders could determine that *great* scores will be greater than or equal to a corrective action impact score of 0.40, *average* scores will be between 0.20 and 0.39, and *basic* scores will be less than 0.20. The groupings will vary from agency to agency. The important factor is for agencies to develop discrete boundaries between categories to facilitate the prioritization process.

Figure 4-6 and Figure 4-7 illustrate how the columns respectively labeled Compliance Gap Percentage and Corrective Action Impact calculations are used at the enterprise level. Sorted by corrective action impact, Figure 4-6 illustrates the ranking process. From left to right, at the enterprise level, the 17 NIST SP 800-26 topic areas are listed, followed by a delineation of the security control areas by agency executives into one of three categories – high (H), moderate (M), or low (L). For example, the first topic area listed—Topic Area 4, Incident Response Capability—received an importance ranking of high.

The compliance gaps are then divided by the cost to implement the corrective action, resulting in the corrective action impact. For the Incident Response Capability, the calculation appears as follows:

$$\left(\frac{65\ \%}{\$81,161} \right) \times 100,000 = .80$$

Figure 4-6 is sorted by corrective action impact because the impact is the focal point of the analysis. The corrective action impact yields a "best bang-for-the buck" proportion that is essential in the prioritization process because it signifies high-impact, low-cost corrective actions.

Finally, the far-right column contains the corrective action impact ranking category. In this example, stakeholders ranked corrective action impacts greater than 0.40 as "great," impacts between 0.20 and 0.39 as "average," and impacts less than 0.20 as "basic." Therefore, Incident Response Capability received a ranking of "great."

	NIST SP 800-26 Topic Area	Importance Ranking	System X	System Y	System Z	Average Across Agency	Security Compliance Gap %	Corrective Action Cost	Corrective Action Impact	Category
4	Incident Response Capability	H	50 %	25 %	30 %	35 %	65 %	$81,161	0.80	G
3	Life Cycle	M	25 %	12 %	15 %	17 %	83 %	$117,789	0.70	G
11	Audit Trails	M	0 %	100 %	75 %	58 %	42 %	$94,326	0.44	G
8	Authorize Processing	H	5 %	10 %	5 %	7 %	93 %	$237,350	0.39	A
10	Security Awareness, Training, and Education	L	50 %	0 %	100 %	50 %	50 %	$133,898	0.37	A
9	Physical Security	M	75 %	75 %	75 %	75 %	25 %	$88,762	0.28	A
14	Logical Access Controls	L	25 %	10 %	75 %	37 %	63 %	$248,154	0.26	A
16	Data Integrity	H	50 %	10 %	0 %	20 %	80 %	$328,506	0.24	A
15	Review of Security Controls	H	75 %	90 %	10 %	58 %	42 %	$179,139	0.23	A
13	Documentation	L	75 %	100 %	50 %	75 %	25 %	$144,755	0.17	B
17	Production, Input/Output Controls	L	25 %	30 %	10 %	22 %	78 %	$457,120	0.17	B
2	Hardware and System Software Maintenance	M	25 %	25 %	25 %	25 %	75 %	$450,959	0.17	B
7	Contingency Planning	M	15 %	25 %	50 %	30 %	70 %	$482,347	0.15	B
12	Risk Management	H	90 %	80 %	65 %	78 %	22 %	$168,335	0.13	B
1	Identification and Authentication	M	50 %	25 %	60 %	45 %	55 %	$441,880	0.12	B
6	Personal Security	M	90 %	90 %	100 %	93 %	7 %	$82,263	0.08	B
5	System Security Plan	M	100 %	95 %	90 %	95 %	5 %	$196,143	0.03	B

Figure 4-6. Enterprise-Level Prioritization[18]

---

[18] The data contained within Figure 4-5 and all other figures within this guidance are for illustrative purposes only. The purpose of this figure is neither to imply that it represents an appropriate ranking of topic areas nor that these are correct importance rankings. These rankings are variables that will differ within the context of each agency.

Based on the analysis in Figure 4-6, the agency is prepared to prioritize its enterprise-level security controls. Figure 4-7 illustrates the 3x3 matrix technique that can be used to prioritize at the enterprise level. If prioritizing via a 3x3 matrix is not viable, a spreadsheet sort by the "Category" and "Ranking" columns would also provide a rank-ordered list that approximates the prioritization achieved through the 3x3 matrix exercise.

Using the findings in Figure 4-6, agencies can plot the 17 NIST SP 800-26 security controls within the 3x3 matrix in Figure 4-7. For example, in Figure 4-6, Topic Area 4, Incident Response Capability, has an importance ranking of "H" for high and a corrective action impact ranking of "G" for great. Therefore, this topic area would be

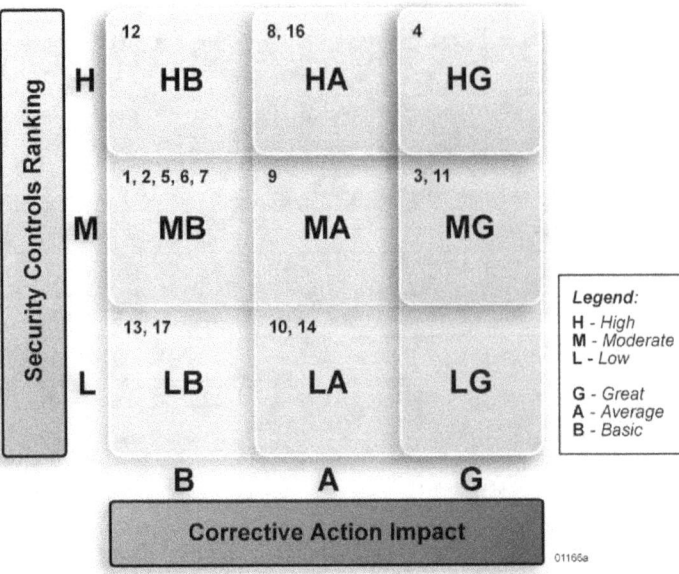

Figure 4-7. Enterprise-Level Prioritization Analysis Matrix

plotted in cell HG, as referenced by the "4" in cell HG. Following this methodology, agencies can plot all security controls within the 3x3 matrix.

The corrective actions mapped to the HG cell will receive the highest priority because they represent the corrective actions that received the highest executive ranking for importance to the agency and provide the greatest reduction in security compliance gaps. In addition, because of the corrective action impact calculation, the corrective actions that align in HG will demonstrate the most cost-effective corrective actions. Corrective action priority then moves from high to low along a diagonal line from the upper-rightmost cell to the lower-leftmost cell. Agency executives must determine the relative importance of peripheral cells HA, MG, HB, LG, MB, and LA. Based on agency goals and stakeholder priorities, agencies need to determine if the corrective action impact axis is more important and weight cell MG higher, or if the security controls ranking axis is more important and weight cell HA higher.

In the example in Figure 4-7, the agency will use a five-step prioritization determination.

1. Topic Area 4, Incident Response Capability, will receive the highest priority for funding because it is the only control area listed in cell HG.

2. The agency will need to determine the relative importance of cells HA and MG to continue the prioritization sequence. If the NIST SP 800-26 security control areas axis is more important, then the HA topic areas will precede the MG topic areas. If the corrective action impact axis is more important, then MG topic areas will precede HA topic areas.

3. The next tier to consider consists of cells HB, MA, and LG. Based on the analysis in step 2, the next cell within this tier will depend on whether the agency stakeholders placed more importance on the NIST SP 800-26 topic areas axis or the corrective action impact axis.

4. The third tier to consider consists of cells MB and LA. Based on the analysis in step 2, the next cell within this tier will depend on whether the agency stakeholders placed more importance on the NIST SP 800-26 topic areas axis or the corrective action impact axis.

5. The prioritization process would conclude with the two topic areas in cell LB.

Following executive validation, the agency should implement the corrective actions from high to low priority across the agency. However, the agency also has corrective actions for systems to consider in addition to enterprise-level needs. The same methodology can be applied to a system prioritization that can then be merged with the enterprise-level prioritization. The ultimate result will provide a prioritized list of the corrective actions to be implemented within an available budget from both prioritizations: enterprise-level and system-level.

4.4.2 System-Level Prioritization

Continuing with the example described in Section 4.4.1, the agency would need to prioritize its systems to determine which ones are the most sensitive and require immediate remediation of their vulnerabilities. The first input for system-level prioritization is system sensitivity.[19]

System sensitivity is usually documented in the investment's system security plan and takes into account the systems' confidentiality, integrity, and availability of data. Based on these three factors, the system's sensitivity is delineated into three categories—high, moderate, and low. This sensitivity categorization can then be used as a prioritization criterion for mitigating corrective actions by identifying the investments with the highest sensitivity.

System category or criticality can be used as a system-level prioritization criterion for the purposes of this guidance. FIPS 199, *Standards for Security Categorization of Federal Information and Information Systems*, contains information on determining a system's criticality. As explained in FIPS 199, system criticality is a factor of the confidentiality, integrity, and availability of the system's information.

The next input is the security compliance percentage. This percentage is obtained by evaluating each system's compliance with the 17 NIST SP 800-26 topic areas. For example, if a system were 60 percent compliant with the 17 NIST SP 800-26 topic areas, then the security compliance percentage would be 60 percent. This figure is then subtracted from 100 percent to yield the security compliance gap. Following the example of a 60 percent security compliance percentage, the security compliance gap would be 40 percent.

The corrective action cost is the next input to the analysis. This number should come directly from the system's POA&M. It would reflect the total corrective action costs to mitigate all identified weaknesses in the POA&M.

Once the security compliance gap and the corrective action cost have been determined, the corrective action impact can be calculated. Using the same formula as the enterprise-level prioritization, the corrective action impact is calculated as follows:

$$\left(\frac{\text{Corrective Action Security Compliance Gap \%}}{\text{Corrective Action Cost}} \right) \times 100{,}000$$

Finally, after the corrective action impact has been calculated, the systems should be prioritized accordingly. The impacts should be delineated into three categories: great, average, and basic. Agency stakeholders from line operations to agency executives should determine the thresholds for each of the three categories. For example, agency stakeholders could determine that *great* scores will be greater than or equal to 10.00, *average* scores will be between 1.00 and 9.99, and *basic* scores will be less than 1.00. The groupings will vary from agency to agency. The important factor is for agencies to develop discrete boundaries between categories to facilitate the prioritization process.

---

[19] For additional information on determining system sensitivity or criticality, reference NIST SP 800-37, *Guide for the Security Certification and Accreditation of Federal Information Systems*; NIST SP 800-18, *Guide for Developing Security Plans for Information Technology Systems*; NIST SP 800-59, *Guideline for Identifying an Information System as a National Security System*; NIST SP 800-60, *Guide for Mapping Types of Information and Information Systems to Security Categories*; and FIPS 199, *Standards for Security Categorization of Federal Information and Information Systems*.

illustrates the system-level prioritization. Using system "N" as an example, moving from left to right, its system security plan indicates that it is a "high" sensitivity system. System N is 15 percent compliant with the 17 NIST SP 800-26 topic areas, yielding an 85 percent security compliance gap. The overall corrective action cost identified in the POA&M is $3,800 to mitigate all vulnerabilities. Based on these inputs, the corrective action impact is calculated as follows:

$$\left(\frac{85\%}{\$3,800}\right) \times 100,000 = 22.49$$

Figure 4-8 is sorted by corrective action impact because it is the focal point of the analysis. The corrective action impact yields a "best bang-for-the-buck" proportion that is essential in the prioritization process because it signifies high-impact, low-cost corrective actions.

System Name	Sensitivity Ranking	Security Compliance Percentage	Security Compliance Gap	Corrective Action Cost	Corrective Action Impact	Category
N	H	15 %	85 %	$3,800	22.49	G
S	M	84 %	16 %	$1,456	11.16	G
F	H	90 %	10 %	$1,000	10.00	G
J	L	12 %	88 %	$17,431	5.02	A
I	H	11 %	89 %	$26,387	3.37	A
K	L	5 %	95 %	$45,566	2.08	A
A	L	10 %	90 %	$75,000	1.20	A
P	M	76 %	24 %	$27,248	0.89	B
H	M	41 %	59 %	$71,860	0.82	B
C	M	25 %	75 %	$100,000	0.75	B
E	H	50 %	50 %	$100,000	0.50	B
G	L	62 %	38 %	$77,954	0.49	B
B	L	40 %	60 %	$125,000	0.48	B
D	L	55 %	45 %	$99,000	0.45	B
M	H	60 %	40 %	$92,423	0.44	B
T	H	78 %	22 %	$53,830	0.40	B
R	M	97 %	3 %	$26,442	0.12	B
O	M	86 %	14 %	$119,060	0.12	B
Q	M	95 %	5 %	$69,627	0.06	B
L	M	99 %	1 %	$31,627	0.05	B

Figure 4-8. System-Level Prioritization

As with the previously discussed enterprise-level prioritization process, a 3x3 matrix can be used to prioritize agency systems. Figure 4-9 shows such a matrix.

The prioritization methodology at the system level mirrors the process at the enterprise level. Systems map to one of the nine cells based on their system sensitivity and corrective action impact scores. For example, system N is a *high* sensitivity system with a *great* corrective action impact. Therefore, system N would map to cell HG. As with the enterprise-level prioritization, the systems that are mapped to the HG cell will receive the highest priority because they represent systems with the highest sensitivity ranking and provide the greatest corrective action impact. The prioritization process would continue in order of importance from the upper-rightmost cell of the matrix on a diagonal line to the lower-leftmost cell of the matrix.

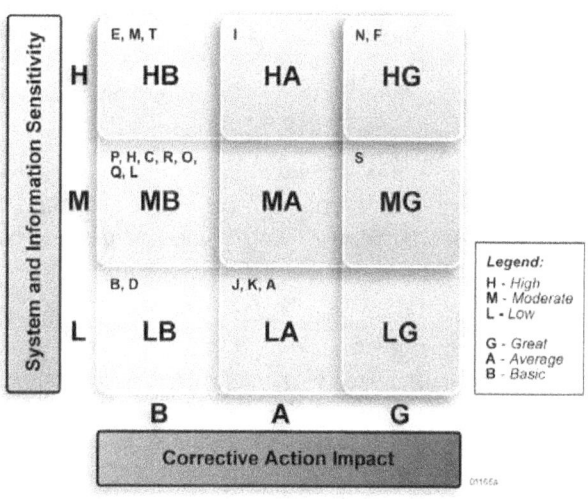

Figure 4-9. System-level Prioritization Analysis Matrix

4.4.3 Joint Prioritization

The final step in the prioritization process is to combine the enterprise- and system-level prioritizations into one prioritization framework to create a security investment strategy for the agency. The agency should map the systems and enterprise-level controls already prioritized in Sections 4.4.1 and 4.4.2 to the 3x3 matrix presented in Figure 4-10. For example, all enterprise-level controls that were in cell HG in Figure 4-7 and all systems that were in cell HG in Figure 4-9 should appear in cell HG in Figure 4-10.

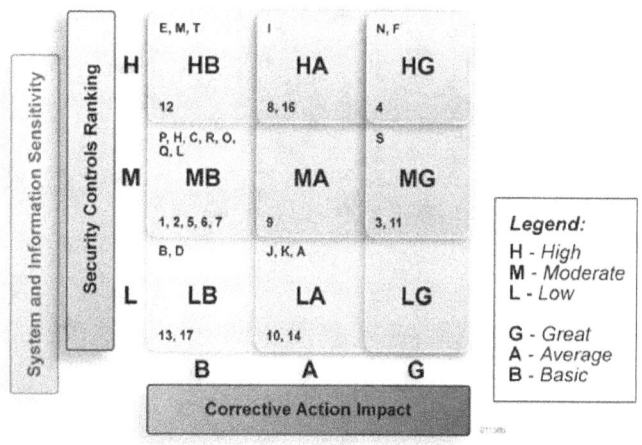

Figure 4-10. Joint Prioritization Analysis Matrix

The prioritization process in Figure 4-10 is conducted in exactly the same manner as the enterprise and system levels. Cell HG represents the highest priority systems and enterprise controls, while cell LB represents the lowest priority systems and enterprise controls. Implementation of enterprise and system-level corrective actions should begin in cell HG and proceed in a diagonal line towards cell LB. The extent of implementation depends on the security budget allocations the agency receives or expects.

For example, adding the costs of all enterprise- and system-level corrective actions (presented in Figure 4-6 and Figure 4-8) across the agency yields a total corrective action cost of $4,840,505. However, if the agency can only allocate $2,000,000 in a particular year to IT security, it can use the prioritization in Figure 4-11 to determine the appropriate allocation of limited security funds.

Figure 4-11 displays the corrective action costs for each cell. These costs were computed by adding the system and enterprise costs for the system- and enterprise-level corrective actions within the cells from Figure 4-6 and Figure 4-8. Once the costs for all prioritized corrective actions are included in the analysis, key management personnel should collectively agree on the prioritization approach and ensure that it aligns with agency priorities and spending plans. For example, in some cases, it might be more cost effective to pursue lesser priority items because of per-seat discounts and other cost/benefit criteria that are outside the scope of the methodology presented in this guidance.

Plotting all of the systems within the agency's portfolio on such a graph will provide a snapshot of the agency's high-sensitivity, high-corrective action impact systems. After plotting all of its systems, the agency should perform executive validation of the placement of the various systems to ensure that stakeholders' priorities are met.

Assuming the agency stakeholders agree that all prioritized corrective actions are appropriate as displayed in Figure 4-10, the analysis can proceed accordingly. As Figure 4-11 demonstrates, adding the three highest priority cells together (HG, HA, and MG) brings a total of $891,775, which is nearly half of the corrective action budget of $2,000,000. The agency would then move to the next tier of prioritization, or cells HB, MA, and LG. Totaling these cells yields a total of $503,350, which combined with the previous total from HG, HA, and MG, yields a running total of $1,395,125.

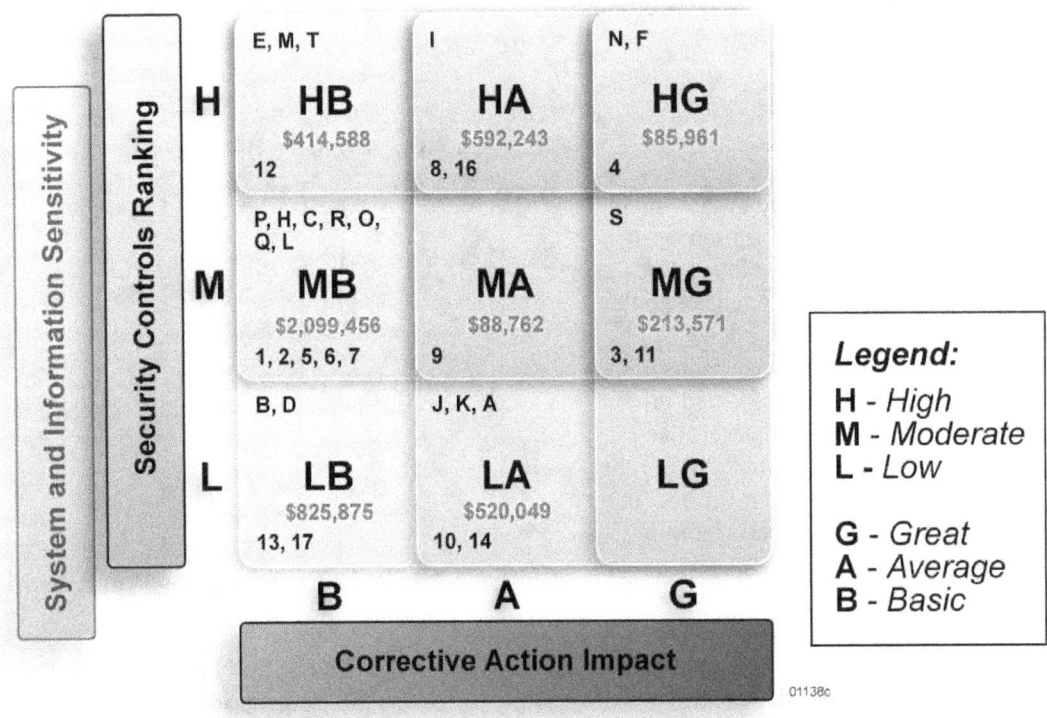

Figure 4-11. Corrective Action Prioritization with Costs

With $604,875 remaining in the corrective action budget, the agency would proceed with prioritization into cells MB and LA. Totaling those two cells yields $2,619,505. Clearly, this total exceeds the remaining corrective action budget, so stakeholders will have to decide on how to allocate the remaining dollars. Should the stakeholders determine that corrective action impact should be the driving factor, the corrective actions in cell LA would be implemented. Should the stakeholders determine that system sensitivity and security control ranking are the driving factors, then selected corrective actions from cell MB would be implemented until the remaining $604,875 is expended.

One benefit of this methodology is that once all corrective action dollars have been expended, the remaining corrective actions are still prioritized according to cost and impact criteria. Therefore, they can be addressed in order of priority during the subsequent budget cycle. Furthermore, implementing enterprise-level corrective actions early in the process could mitigate system-level weaknesses. For example, if an enterprise corrective action calls for all IT systems to be certified and accredited, that corrective action would mitigate any C&A-related vulnerabilities among the system-level weaknesses. Therefore, dollars would not have to be allocated to the particular systems with C&A weaknesses and could be allocated elsewhere according to the prioritization framework.

> In some instances, agencies may find that available funding does not cover mitigation of all of the prioritized vulnerabilities that the agency would like to fund for the budget year. In these instances, the agency should evaluate corrective actions that fall below the cut line and pursue a variety of strategies to fund the corrective actions. Agencies can look into shifting resources throughout the year or reprogramming other dollars to fund corrective actions below the cutoff line.

Regardless of budget constraints or the number of enterprise- and system-level corrective actions, by using this model, agencies will have an easily updateable roadmap for corrective action implementation that provides an action plan for mitigating risks.

4.5 Develop Supporting Materials

Once prioritizing against requirements is completed, operating units are poised to select their investments for the budget year and begin the process of requesting funding from OMB for the next year to implement the corrective actions and security controls.

4.5.1 Enterprise- and System-Level Considerations

Prioritized enterprise- and system-level corrective actions become candidates for investment. While the supporting materials process is essentially the same for each type of investment, requirements and drivers differ, as shown in Figure 4-12.

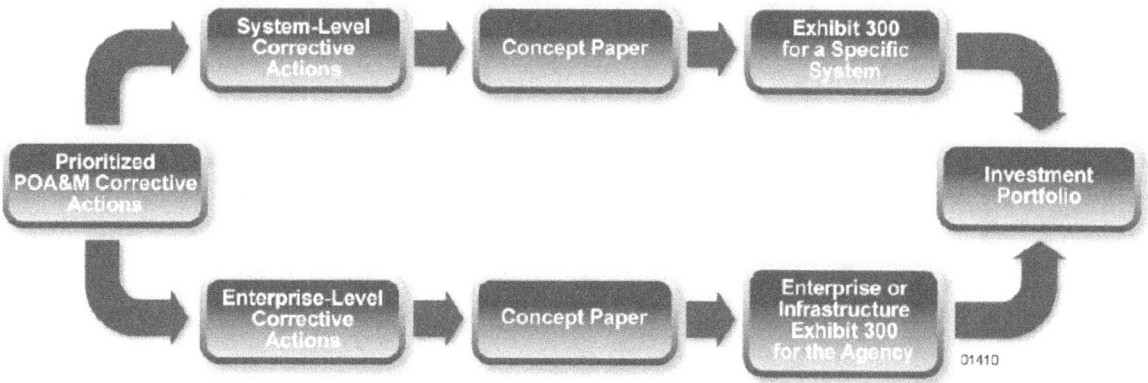

Figure 4-12. Enterprise- and System-Level Requirements

Once POA&M corrective actions are prioritized, system-level corrective actions will require system-level documentation, while enterprise-level corrective actions will require enterprise- or infrastructure-level documentation.

4.5.2 Concept Paper

Regardless of whether the potential investment is at the system or enterprise/infrastructure level, the suggested budget process for new investments begins with a concept paper. The concept paper is developed by the investment owner and submitted to the IRB for review. The concept paper provides a high-level description of the proposed investment and includes a rough-order-of-magnitude costing estimate, benefits, milestones, and agency impacts. Such papers are usually only a few pages long. Based on the concept paper, the IRB can determine whether the investment will be a worthwhile endeavor and recommend continuation or cancellation of the potential investment.

4.5.3 Investment Thresholds

Following approval of the concept paper, the Select phase of the investment life cycle begins. The investment will require additional budget documentation for internal (IRB, TRB, *etc.*) and external (OMB) review. The degree of formality of required documentation is commensurate with the investment thresholds. Figure 4-13 presents sample IT security investment review thresholds for illustrative purposes only. Each agency will have its own thresholds, depending on agency mission and overall budget. Existence of formal investment thresholds that would trigger a more rigorous level of review are indicative of agencies that have moved from Stage 1 to Stage 2 of the ITIM maturity model.

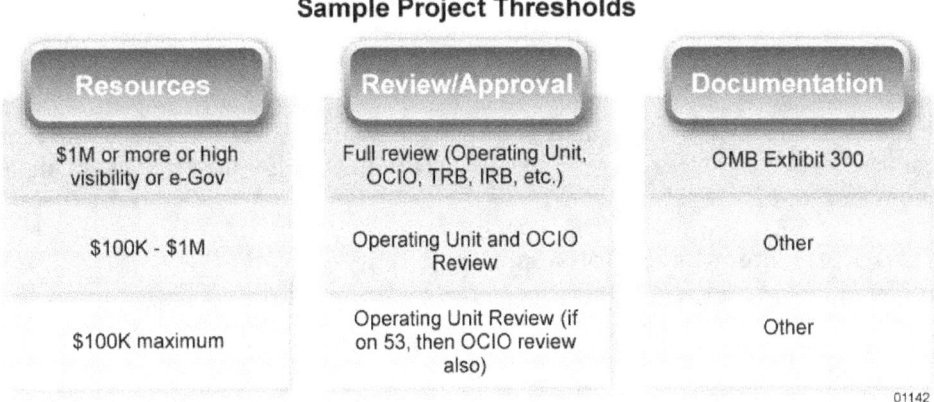

Figure 4-13. Illustrative Project Thresholds

As Figure 4-13 demonstrates, the greater the life-cycle cost of the security investment, the more rigorous the review process. Generally, operating units, with approval from the OCIO, use their discretion when funding investments are below $1 million. However, for investments that are e-Gov, high profile, or over $1 million, a full review by the operating unit, TRB, IRB, and OCIO is necessary to demonstrate that all requirements are met and that the investment aligns with agency mission.

During this time, the investment owner must complete a series of assessments and activities to ensure formal planning and development takes place and that all requirements are met. These activities are detailed in Figure 4-14.

4.5.4 The Exhibit 300

As Figure 4-14 illustrates, the Exhibit 300 is the capture mechanism for all of the analyses and activities required for full internal (IRB, OCIO) review. More importantly, the Exhibit 300 is the document that OMB uses to assess investments and ultimately make funding decisions. The Exhibit 300 also provides OMB with a robust assessment of the investment and is the vehicle for IT investments to justify life-cycle and annual funding requests to OMB. The Exhibit 300 will:

- Provide a means for planning, budgeting, and acquiring capital assets
- Provide a technical basis and defense for investments
- Establish a clear baseline against which progress can be measured

- Document the planning performed for a capital investment.

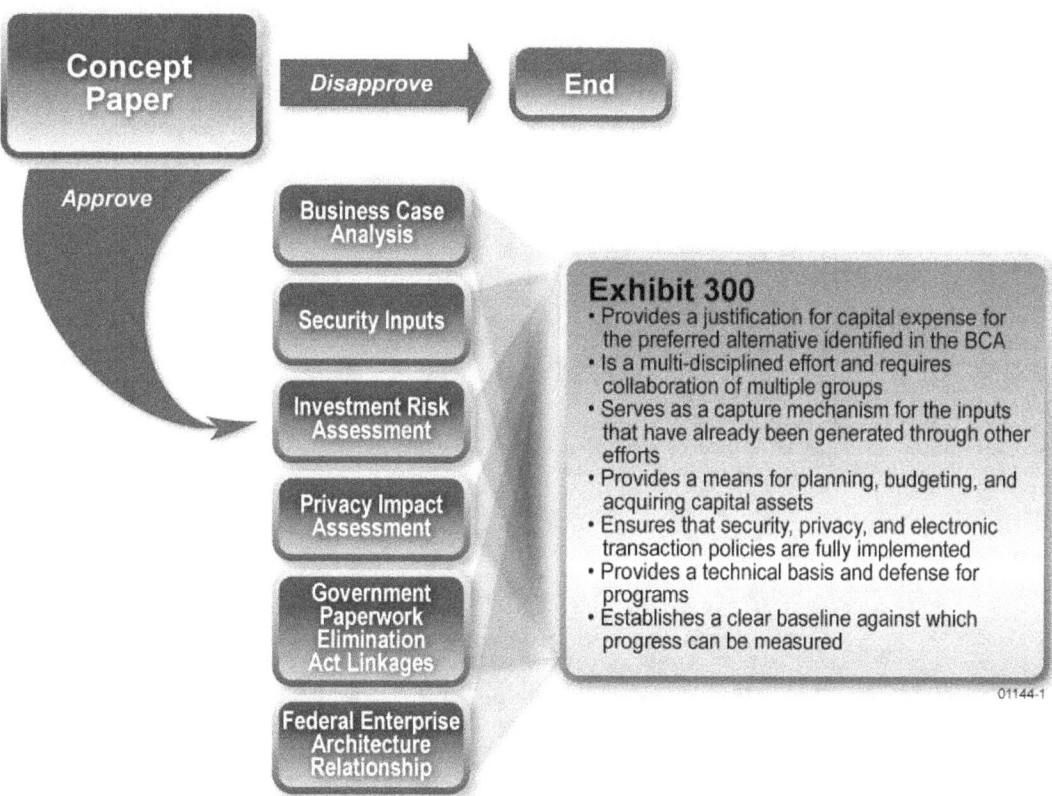

Figure 4-14. The Investment Process, Culminating in the Exhibit 300

The Exhibit 300 is completed for new IT investments and is resubmitted annually for mixed life-cycle and steady-state investments. Operating units should evaluate their prioritized corrective actions and security controls identified during the prioritization process and determine whether the outputs need to be incorporated into an existing investment's Exhibit 300, or whether they will need to create an independent Exhibit 300 for a new investment. For example, if a prioritized corrective action were to implement stronger password protection on System X, then the corrective action would be included as a supplemental budget request under System X's Exhibit 300. However, if a prioritized corrective action were to purchase an automated C&A tool to improve C&A across the agency, then this IT investment would be a new investment and require a concept paper and a separate Exhibit 300 for a new acquisition.

Table 4-1 details the contents of each of the twelve sections within the two parts of the Exhibit 300:[20]

---

[20] Based on 2004 guidance for FY06 Exhibit 300s. For more information on the current Exhibit 300, see the OMB web site at: http://www.whitehouse.gov/omb/circulars/a11/current_year/s300.pdf.

Table 4-1. Exhibit 300 Requirements

Section	Description
Part I: Capital Asset Plan and Business Case (All Assets)	
Summary of Spending	A table detailing historical, current year, budget year, and future planned spending for investment planning, full acquisition, and maintenance activities as well as government full-time equivalent costs
Project Description	A description of the investment, including its status in the agency's CPIC "control" review process, any noteworthy assumptions made about the project, and any other supporting documentation
Justification	• A demonstration of investment alignment with the PMA and agency mission and goals • Explanations of stakeholders, customers and any collaboration with other agencies • Explanation of why other private/public sector alternatives are not viable • Descriptions of any linkages with other investments or agencies • Descriptions of any benefits or cost reductions
Performance Goals and Measures	A table detailing specific, measurable performance goals, using the FEA Performance Reference Model (PRM)
Program Management	A description of investment project management structures, responsibilities, and qualifications that contribute to successful achievement of cost, schedule, and performance goals
Alternatives Analysis	A description of the costs, benefits, and risks of three viable alternatives that were compared consistently, and a detailed description of the preferred alternative, including selection rationale
Risk Inventory and Assessment	A discussion of the 19 OMB-defined areas of risk as they apply to the particular investment (referenced in Section 2.6), including plans to mitigate and/or manage the risk
Acquisition Strategy	A description of the investment's acquisition strategy and a discussion of how it mitigates risk to the federal government, accommodates Section 508 as needed, and uses performance-based contracts
Project and Funding Plan	The use of an EVMS and/or operational assessments to illustrate how the investment is meeting its cost, schedule, and performance goals
Part II: Additional Business Case Criteria for IT	
Enterprise Architecture	A discussion of how the investment: • Supports the FEA • Is integrated within the agency's EA • Shows how the business, data, and technology layers of the EA relate to the investment
Security and Privacy	A demonstration that • Thorough security and privacy planning have taken place • Security has been accounted for in the investment's life-cycle costs • The investment is fully certified and accredited • A privacy impact assessment has been completed for this investment
Government Paperwork Elimination Act (GPEA)	A description of how the investment relates to the agency's GPEA plan and whether it supports electronic transactions or record keeping that is covered by GPEA

The Alternatives Analysis section of the Exhibit 300 is a key step in making sound business decisions, especially in the IT security arena. Every Exhibit 300 must include a minimum of three alternatives, which should demonstrate:

- A way of meeting the mission need or providing functionality needed to accomplish mission/goals (*e.g.*, sometimes a mission need can be met with a new IT system; sometimes by changing business processes)
- Why the selected investment provides the most effective (cost- and performance-wise) manner of meeting the associated mission need (versus different investments or process changes).

This section is critical for enterprise-level security investments because investment owners must demonstrate why the selected security solutions are the most effective. The objective, quantified prioritization process discussed in Section 4.4 links security decisions to agency goals and investment impact, which can be used to justify alternatives.

Some examples of credible alternatives include:

- **Status Quo.** Status quo, or an explanation of the current method of meeting the mission need, should always be one of the alternatives. This alternative explains the limitations and/or adverse effects on performance associated with the current status. Presumably, an investment would be needed because the current way of meeting the mission need is inadequate.
- **Outsourcing.** This alternative will analyze and document benefits, risks, and costs of outsourcing the function.[21]
- **Government-Owned and -Operated.** This alternative will analyze and document benefits, risks, and costs of maintaining the function within and ownership of assets by the government.
- **Process/Organizational Changes Only.** This alternative could include a reorganization in the agency or division or the reengineering of a particular business process that helps an organization meet a mission need. This alternative will also analyze and document benefits, risks, and costs of restructuring processes or functions within the agency versus meeting the mission need with the investment.
- **IT/System Only.** This alternative involves investing in a system or an IT asset without any underlying organizational changes.

Once alternatives are selected, the input assessments and activities are completed, and the sections of the Exhibit 300 are finalized, the operating unit's Exhibit 300s are forwarded to the agency's IRB for review.

4.6 IRB and Portfolio Management

The IRB reviews and selects investments based on the Exhibit 300s forwarded by operating units. Like the prioritization that occurs at the operating unit level, the IRB typically uses strategic selection criteria to rank-order the investment pool and usually makes decisions based on agency mission and goals, not just on cost. Security typically is not the driving force behind portfolio management. However, it is strategically important for the investment strategy because it serves as a qualifier for receiving funding and as a business enabler for those functions that cannot be performed without appropriate security controls.

After prioritizing and approving select Exhibit 300s, the IRB forms an investment portfolio request for review by OMB.

4.7 Exhibits 53, 300, and Program Management

Following selection into the agency's IT portfolio, Exhibit 300s are rolled into the Exhibit 53. The Exhibit 53 provides an overview of the agency's entire IT portfolio by listing every IT investment, life cycle, and budget-year cost information. The Exhibit 53 has four sections:

1. IT systems by mission area
2. IT infrastructure and office automation
3. EA and planning
4. Grants management.

In addition to containing all investments with Exhibit 300s, the Exhibit 53 also contains other IT investments that do not have Exhibit 300s (for example, legacy systems with costs below agency thresholds).

---

[21] For more information on outsourcing and IT services, see NIST SP 800-35, *Guide to Information Technology Security Services*.

OMB evaluates an agency's Exhibit 53 and Exhibit 300s and determines appropriate funding amounts for the budget year based on the justification articulated in the Exhibit 300s. Agencies then receive their budget year funding and must implement or maintain their investments throughout the year by applying allocated funding.

For investments in the Control and Evaluate phases of the investment life cycle, project managers must manage their investments and demonstrate progress against the baseline in the Exhibit 300 annually to continue receiving funding.

5. Implementation Issues

After prioritizing IT security investments and developing Exhibit 300s and the Exhibit 53, the agency must implement and monitor these investments. Throughout the implementation process, IT security decisions are made based on system security issues and federal budgeting timelines.

5.1 IT Security Organizational Processes

After the IRB selects the appropriate investment mix for the IT portfolio and OMB issues funding allocations, the IT security investments must be managed, monitored, and reported on throughout the budget year. This IT security integration cascades throughout three organizational levels, as depicted in Figure 5-1.

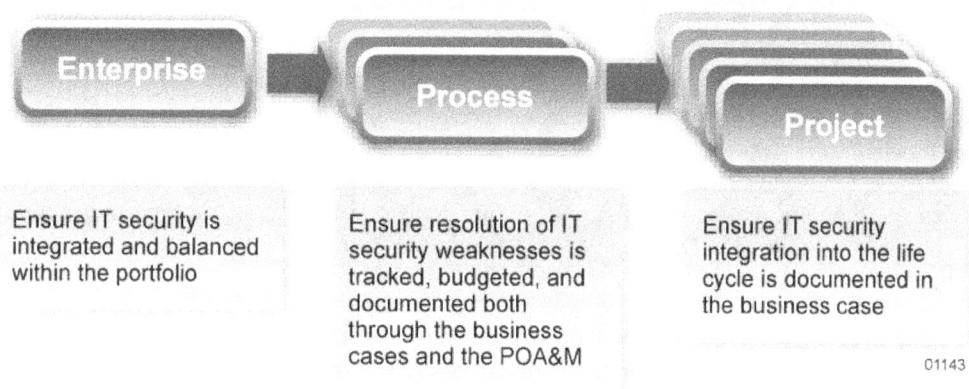

Figure 5-1. Layers of Integration of Security into the CPIC Process

IT security integration begins at the enterprise level, where agency executives must ensure that IT security is integrated throughout the organization. Agency security policy must be developed and supporting procedures implemented to ensure a secure operating environment with tolerable residual risk consistent with federal standards and legislation. This includes application of security controls commensurate with assessment of risk.

Stemming from the organizational level, processes throughout the organization examine security posture and compliance with mandates and legislation, including FISMA. At the process level, agencies need to ensure that IT security weaknesses are continually identified, tracked, budgeted for, and mitigated. The POA&M is the primary vehicle for tracking IT security weaknesses. Within the POA&M, weaknesses and mitigation resources are identified, resolution milestones are established, and weaknesses are tracked until mitigation.

At the project level, IT security project managers should use processes such as the POA&M, FISMA reporting, C&A, *etc.*, to account and budget for IT security annually and over the investment life cycle. Business cases and Exhibit 300s should demonstrate life-cycle security costs and identify any increased costs for newly identified weaknesses or for compliance with regulations. The process then cycles back to the enterprise level where the OCIO, IRB, TRB, *etc.*, evaluate Exhibit 300s annually for IT security compliance and integration. Thus, integrating IT security into the CPIC process requires accountability, decision making, and disciplined procedures throughout the various levels of the organization.

5.2 Project Management

As investments continue to move through their life cycle, it is essential that project managers continue to examine the investment's cost, schedule, and performance indicators. For D/M/E investments, EVM, described in Section 2.3, can be used after the investment is selected for inclusion in the agency's IT portfolio to assess the investment's cost, schedule, and performance. By examining established earned value metrics over time, the project manager can determine whether an investment is behind schedule, over budget, or not meeting identified performance targets. Furthermore, earned value measures used during the Control phase of the investment life cycle enable project managers to evaluate investment outputs versus expenditures to monitor the investment's security cost performance as it matures. Using the corrective action impact scores from the prioritization methodology presented in this guidance, agencies should be able to forecast the impact that a corrective action will provide.

If an investment is late, over cost, or not meeting performance expectations, agency senior executives can use the results of the compiled cost, benefit, and performance metrics data to determine if the investment should:

- Continue through the investment management life cycle unchanged
- Be modified and continue
- Be canceled in its entirety.

EVM metrics and their subsequent ramifications on the decision to continue, modify, or cancel the investment should be included in D/M/E investments' annual Exhibit 300 to justify continued or changed funding requests.

For steady-state or operational investments, project managers should conduct operational assessments.[22] The operational assessment is a formal analysis to determine whether the investment is meeting program objectives and the needs of the owners and users, and whether it is performing within baseline cost, schedule, and performance goals. Operational analyses should take place in accordance with a schedule of fixed milestones established during investment planning or on a cyclical basis. The NIST SP 800-26 self-assessment provides a comprehensive review of an investment's IT security controls and can be used to analyze IT security controls for operational assessments for steady-state investments.

5.3 Legacy Systems

Even with the formal EVM and ITIM approaches, legacy systems present a unique set of challenges for agencies. Often, agencies may rank legacy systems low from a prioritization standpoint because they:

- Include existing systems with historically low development/procurement and corrective action costs
- Have accepted residual risk
- Are perceived to be of a limited life span
- Are often well into the IT life cycle and typically at either the operating and maintenance or disposition stages.

However, if legacy systems are still able to show a direct link to agency mission, they should continue to receive funding. Thus, security is still an important consideration for legacy systems. Typically, IT security issues associated with legacy systems include:

- Lack of current security documentation such as security plans and risk analyses
- Insufficient management, technical, or operational controls such as those associated with continuity of operations planning and disaster recovery.

---

[22] Capital Programming Guide Supplement to Part 7 of OMB Circular No. A-11, "Management in Use Phase."

It is imperative that agencies ensure that sufficient funds are budgeted for and that security is sufficiently integrated into these systems commensurate with system sensitivity and risk. IT security planning and implementation should be applied to legacy systems with the same amount of rigor and discipline as is used on new investments.

5.4 Timelines

It is crucial for agency IT security staff to understand the timelines associated with capital plans and the budget cycle. Even though the Exhibit 53 and Exhibit 300s are submitted to OMB each September, the budgeting process is not confined to the late summer months. Planning, acquiring, and executing IT security budgets are year-round activities. Figure 5-2 indicates prior year, current year, budget year, and second budget year activities that occur in parallel processes.

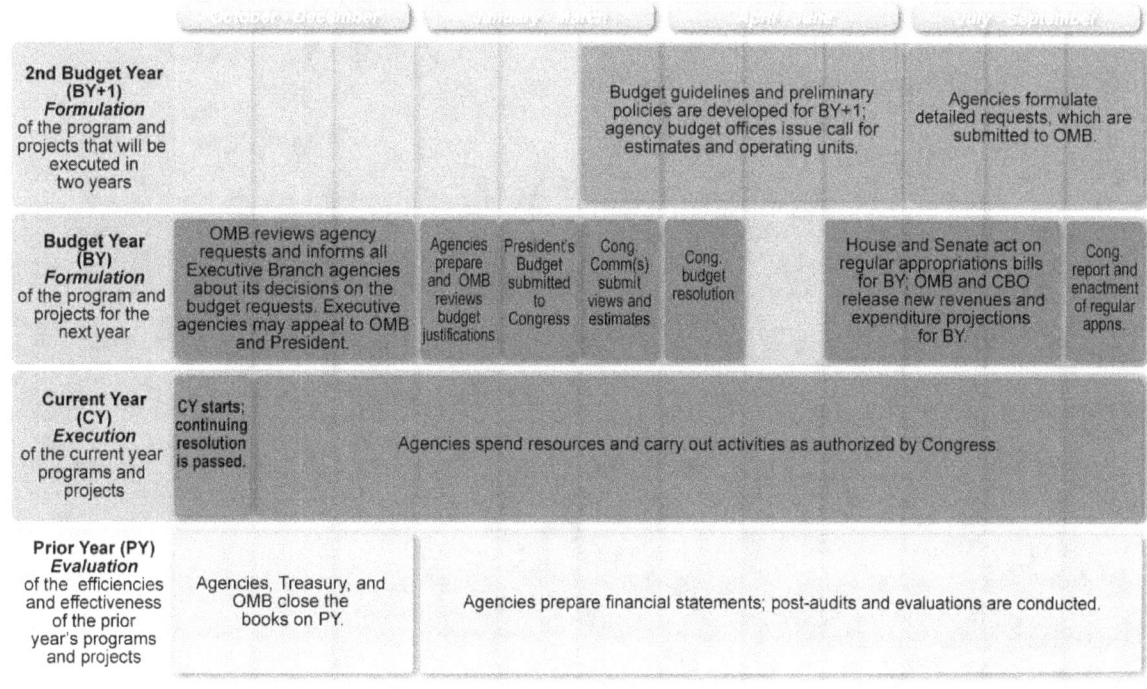

Figure 5-2. Budget Timelines

During the current year, agencies execute their budgets allocated by OMB and Congress. At the same time, agencies evaluate prior year financial and operational performance through audits and evolutions. In addition, while agencies are executing current year budgets, they are planning for the next budget year. Furthermore, agencies begin considering strategies for the second budget year (BY+1) in the current year.

Determinations of current year, prior year, and budget year revolve around October 1, which is the beginning of the government's FY. Beginning October 1, agencies begin to execute the current year budget. For example, on October 1, 2004, agencies would begin to execute the FY05 budget. At the same time, agencies would evaluate their prior year (FY04) budget. Meanwhile, during October, OMB finalizes its review of budget year (FY06) Exhibit 300s submitted by agencies in September. During November and into December, OMB and agencies engage in pass back, whereby OMB returns weak Exhibit 300s to agencies and suggests strategies for improvement. OMB then consolidates the investment requests and budget submissions into a draft Presidential budget that the President receives in January. Upon approval by the President, the budget is later submitted to the House for appropriation.

Figure 5-3 presents the CPIC process point of view for the IT security budgeting timelines. As illustrated in the Figure 5-3, with multiple events of the budget process occurring within each Financial Year, it is imperative that agencies use disciplined CPIC processes and controls to streamline activities.

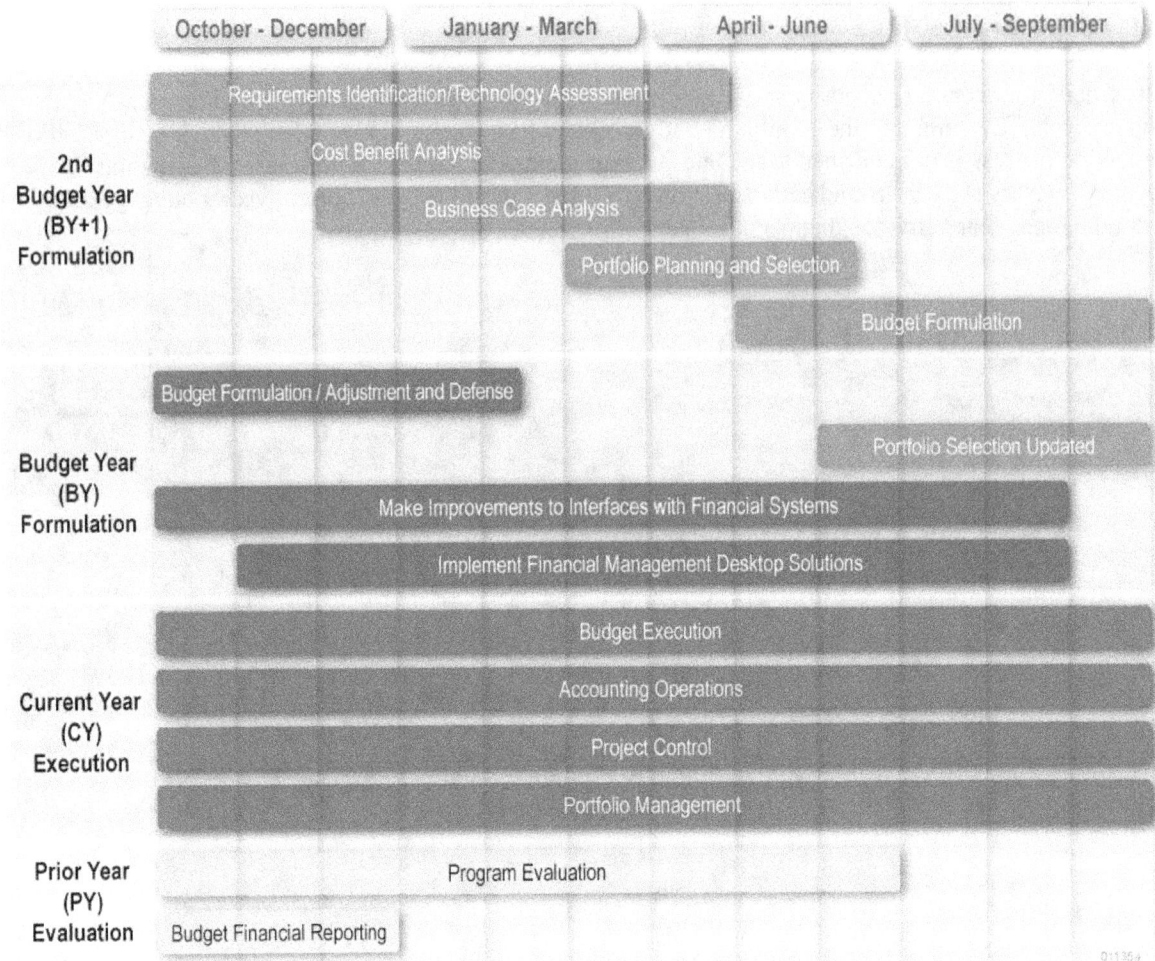

Figure 5-3. CPIC Timelines

Select phase activities performed in the current year are applied to the first and second budget years. During the current year, agencies plan ahead for the two future out-years by identifying potential investments, conducting cost/benefit analyses, developing budgets, and selecting investments to include in the IT investment portfolio.

Control phase activities are performed during the current year as agencies execute their budgets and implement their project controls to ensure schedule and financial milestones are achieved.

Finally, Evaluate phase activities are conducted during the current year for prior year investments to determine whether the investments achieved their intended results.

IT security capital planning is conducted for the future, the present, and the past on a year-round basis. Continual planning, implementation, monitoring, and evaluation leads to a mature CPIC process that not only facilitates improved reporting to OMB and Congress but also leads to an increased security posture and more efficient internal controls and processes.

Appendix A. Glossary[23]

- **Capital planning and investment control (CPIC)** – a synonym for capital programming and is a decision-making process for ensuring that information technology (IT) investments integrate strategic planning, budgeting, procurement, and the management of IT in support of agency missions and business needs. The term comes from the Clinger-Cohen Act of 1996 and generally is used in relationship to IT management issues.

- **The Clinger-Cohen Act of 1996** – legislation that requires agencies to use a disciplined CPIC process to acquire, use, maintain and dispose of information technology.

- **Earned value management (EVM)** – a project (investment) management tool that effectively integrates the investment scope of work with schedule and cost elements for optimum investment planning and control. The qualities and operating characteristics of EVM systems are described in American National Standards Institute (ANSI)/Electronic Industries Alliance (EIA) Standard – 748–1998, Earned Value Management Systems, approved May 19, 1998. It was reaffirmed on August 28, 2002. A copy of Standard 748 is available from Global Engineering Documents (1–800–854–7179). Information on EVM systems is available at www.acq.osd.mil/pm.

- **Federal Enterprise Architecture (FEA)** – a framework that describes the relationship between business functions and the technologies and information that support them. Major IT investments will be aligned against each reference model within the FEA framework.

- **The Federal Information Security Management Act (FISMA)** – requires agencies to integrate IT security into their capital planning and enterprise architecture processes at the agency, conduct annual IT security reviews of all programs and systems, and report the results of those reviews to the Office of Management and Budget (OMB).

- **Government Paperwork Elimination Act (GPEA)** – requires federal agencies to allow individuals or entities that deal with the agencies the option to submit information or transact with the agency electronically, when practicable, and to maintain records electronically, when practicable. The Act specifically states that electronic records and their related electronic signatures are not to be denied legal effect, validity, or enforceability merely because they are in electronic form, and encourages federal government use of a range of electronic signature alternatives.

- **IT security investment** – an IT application or system that is solely devoted to security. For instance, intrusion detection systems (IDS) and public key infrastructure (PKI) are examples of IT security investments.

- **Life-cycle costs** – the overall estimated cost, both Government and contractor, for a particular program alternative over the time period corresponding to the life of the program, including direct and indirect initial costs plus any periodic or continuing costs of operation and maintenance.

- **Major IT investment** – a system or investment that requires special management attention because of its importance to an agency's mission; was a major investment in the previous budget submission and is continuing; is for financial management and spends more than $500,000; is directly tied to the top two layers of the FEA (Services to Citizens and Mode of Delivery); is an integral part of the agency's modernization blueprint (enterprise architecture); has significant program or policy implications; has high executive visibility; and is defined as major by the agency's CPIC process. OMB may work with the agency to declare other investments as major investments. All major investments must be reported on exhibit 53. All major investments must submit a "Capital Asset Plan and Business Case," Exhibit 300. Investments that are e-Government

---

[23] Glossary definitions are adapted from OMB Circular A-11, Part 7, Planning, Budgeting, Acquisition, and Management of Capital Assets, and from applicable NIST guidance.

in nature or use e-business technologies must be identified as major investments regardless of the costs. If unsure about what investments to consider as "major," consult your agency budget officer or OMB representative. Systems not considered "major" are "non-major."

- **Mixed life-cycle investment** – an investment that has both development/modernization/ enhancement (D/M/E) and steady-state aspects. For example, a mixed life-cycle investment could include a prototype or module of a system that is operational with the remainder of the system in D/M/E stages; or, a service contract for steady state on the current system with a D/M/E requirement for system upgrade or replacement.

- **Privacy impact assessment** – a process for examining the risks and ramifications of collecting, maintaining, and disseminating information in identifiable form in an electronic information system, and for identifying and evaluating protections and alternative processes to mitigate the impact to privacy of collecting information in identifiable form. Consistent with September 26, 2003, OMB guidance (M-03-22) implementing the privacy provisions of the e-Government Act, agencies must conduct privacy impact assessments for all new or significantly altered IT investments administering information in identifiable form collected from or about members of the public. Agencies may choose whether to conduct privacy impact assessments for IT investments administering information in identifiable form collected from or about agency employees.

- **Risk**
    - **Security risk** – the level of impact on agency operations (including mission functions, image, or reputation), agency assets, or individuals resulting from the operation of an information system given the potential impact of a threat and the likelihood of that threat occurring.
    - **Investment risk** – risks associated with the potential inability to achieve overall program objectives within defined cost, schedule, and technical constraints. OMB has defined 19 areas of investment risk, all of which are required to be addressed in the Exhibit 300.

- **Select-Control-Evaluate IT investment management process**
    - **Select** – the goal of the Select phase is to assess and prioritize current and proposed IT projects and then create a portfolio of IT projects. In doing so, this phase helps to ensure that the organization (1) selects those IT projects that will best support mission needs and (2) identifies and analyzes a project's risks and returns before spending a significant amount of project funds. A critical element of this phase is that a group of senior executives makes project selection and prioritization decisions based on a consistent set of decision criteria that compares costs, benefits, risks, and potential returns of the various IT projects.
    - **Control** – the Control phase consists of managing investments while monitoring for results. Once the IT projects have been selected, senior executives periodically assess the progress of the projects against their projected cost, scheduled milestones, and expected mission benefits.
    - **Evaluate** – the Evaluate phase provides a mechanism for constantly improving the organization's IT investment process. The goal of this phase is to measure, analyze, and record results based on the data collected throughout each phase. Senior executives assess the degree to which each project has met its planned cost and schedule goals and has fulfilled its projected contribution to the organization's mission. The primary tool in this phase is the post-implementation review (PIR), which should be conducted once a project has been completed. PIRs help senior managers assess whether a project's proposed benefits were achieved and also help to refine the IT selection criteria to be used in the future.

- **Security controls** – the management, operational, and technical controls (*e.g.*, safeguards or countermeasures) prescribed for an information system to protect the confidentiality, integrity, and availability of the system and its information.

- **Steady State** – an asset or part of an asset that has been delivered and is performing the mission.

Appendix B. Acronyms

ACWP	Actual Cost of Work Performed
BCA	Business Case Analysis
BCWP	Budget Cost of Work Performed
BCWS	Budgeted Cost of Work Scheduled
BLSR	Baseline Security Requirement
C&A	Certification and Accreditation
CBSR	Cost, Benefit, Schedule, and Risk
CFO	Chief Financial Officer
CIO	Chief Information Officer
CPIC	Capital Planning Investment Control
D/M/E	Development, Modernization, and/or Enhancement
EA	Enterprise Architecture
EVM	Earned Value Management
EVMS	Earned Value Management System
FEA	Federal Enterprise Architecture
FIPS	Federal Information Processing Standard(s)
FISCAM	Federal Information System Controls Audit Manual
FISMA	Federal Information Security Management Act
FY	Fiscal Year
GAO	Government Accountability Office
GPEA	Government Paperwork Elimination Act
IDS	Intrusion Detection System
IG	Inspector General
IRB	Investment Review Board
ISO	Information Security Officer
IT	Information Technology
ITIM	Information Technology Investment Management
ITL	Information Technology Laboratory
NIST	National Institute of Standards and Technology
OCIO	Office of the Chief Information Officer
OMB	Office of Management and Budget
PIR	Post-Implementation Review
PKI	Public Key Infrastructure
PMA	President's Management Agenda
POA&M	Plan of Action and Milestones
POC	Point of Contact
PY	Prior Year
ROSI	Return on Security Investment
SDLC	System Development Life Cycle
SP	Special Publication
ST&E	Security Test and Evaluation
TRB	Technical Review Board
WBS	Work Breakdown Structure

Appendix C. References

American National Standards Institute (ANSI)/Electronic Industries Alliance (EIA) Standard –748–1998, *Earned Value Management Systems,* May 19, 1998, and reaffirmed on August 28, 2002.

Federal Information Processing Standard 199, *Standards for Security Categorization of Federal Information and Information Systems,* February 2004.

GAO/AIMD-12.19.6, *Federal Information System Controls Audit Manual,* January 1999.

GAO-03-1028, *Information Technology: Departmental Leadership Crucial to Success of Investment Reforms at Interior,* September 12, 2003.

GAO-04-394-G, *Information Technology Investment Management: A Framework for Assessing and Improving Process Maturity,* Version 1.1, March 2004.

NIST Special Publication 800-18, *Guide for Developing Security Plans for Information Technology Systems,* December 1998.

NIST Special Publication 800-26, *Security Self-Assessment Guide for Information Technology Systems,* November 2001.

NIST Special Publication 800-30, *Risk Management Guide for Information Technology Systems,* January 2002.

NIST Special Publication 800-35, *Guide to Information Technology Security Services,* October 2003.

NIST Special Publication 800-37, *Guide for the Security Certification and Accreditation of Federal Information Technology Systems,* May 2004.

NIST SP 800-53, *Recommended Security Controls for Federal Information Systems,* September 2004.

NIST Special Publication 800-55, *Security Metrics Guide for Information Technology Systems,* July 2003.

NIST Special Publication 800-59, *Guideline for Identifying an Information System as a National Security System,* August 2003.

NIST Special Publication 800-60, *Guide for Mapping Types of Information and Information Systems to Security Categories,* September, 2004.

NIST Special Publication 800-64, *Security Considerations in the Information System Development Life Cycle,* October 2003.

OMB Circular A-11, *Preparation, Submission, and Execution of the Budget,* 2003.

OMB Circular A-130, *Management of Federal Information Resources,* 2003.

Appendix D. Security Requirements Mapping

Table D-1. Security Requirements Mapping

OMB A-11	NIST SP 800-26 Topic Area	NIST SP 800-53 Security Control Families	Implementation Guidance[24]
Direct			
Risk Assessment	1. Risk Management	Risk Assessment (RA)	~ NIST SP 800-30, Risk Management Guide for Information Technology Systems
Security Planning and Policy	5. System Security Plan	Planning (PL)	~ NIST SP 800-18, Guide for Developing Security Plans for Information Technology Systems
Certification and Accreditation (C&A)	4. Authorize Processing	Certification, Accreditation, and Security Assessments (CA)	~ NIST SP 800-37, Guidelines for the Security Certification and Accreditation of Federal Information Technology Systems ~ NIST SP 800-53, Recommended Security Controls for Federal Information Systems
Specific Management, Operational, and Technical Security Controls	11. Data Integrity 16. Logical Access Controls	Access Control (AC) System and Information Integrity (SI)	~ NIST SP 800-53, Recommended Security Controls for Federal Information Systems ~ NIST SP 800-26, Security Self-Assessment Guide for Information Technology Systems
Authentication or Cryptographic Applications	15. Identification and Authentication	System and Communications Protection (SC) Identification and Authentication (IA)	~ NIST SP 800-21, Guideline for Implementing Cryptography in the Federal Government ~ NIST SP 800-25, Federal Agency Use of Public Key Technology for Digital Signatures and Authentication ~ NIST SP 800-63, e-authentication ~ FIPS 140-2, Security Requirements for Cryptographic Modules ~ FIPS 201, Personal Identification Verification for Federal Employees and Contractors
Education, Awareness, and Training	13. Security Awareness, Training, and Education	Awareness and Training (AT)	~ NIST SP 800-16, Information Technology Security Training Requirements: A Role and Performance-Based Model ~ NIST SP 800-50, Building an Information Technology Security Awareness and Training Program
System Reviews/ Evaluations (includes ST&E)	2. Review of Security Controls	Certification, Accreditation, and Security Assessments (CA)	~ NIST SP 800-53, Recommended Security Controls for Federal Information Systems ~ NIST SP 800-26, Security Self-Assessment Guide for Information Technology Systems

---

[24] For the current version and status of each guide, visit http://csrc.nist.gov/publications/nistpubs/index.html. NIST SP 800-53 addresses all components listed above and is not redundantly listed in the column "Implementation Guidance".

OMB A-11	NIST SP 800-26 Topic Area	NIST SP 800-53 Security Control Families	Implementation Guidance[a]
Oversight or Compliance Inspections		Audit and Accountability (AU)	~ NIST SP 800-53, Recommended Security Controls for Federal Information Systems ~ NIST SP 800-26, Security Self-Assessment Guide for Information Technology Systems ~NIST SP 800-35, Guide to Information Technology Security Services ~ NIST SP 800-18, Guide for Developing Security Plans for Information Technology Systems
Development or Maintenance of Agency Reports to OMB and Corrective Action Plans as They Pertain to the Specific Investment	3. Life Cycle 2. Review of Security Controls	Certification, Accreditation, and Security Assessments (CA) Planning (PL)	~ OMB FISMA Reporting Guidance ~ NIST SP 800-55, Security Metrics Guide for Information Technology Systems ~ NIST SP 800-64, Security Considerations in the Information System Development Life Cycle
Contingency Planning and Testing	9. Contingency Planning	Contingency Planning (CP)	~ NIST SP 800-34, Contingency Planning Guide for Information Technology Systems
Physical and Environmental Controls for Hardware and Software	8. Production, Input/Output Controls	Physical and Environmental Protection (PE)	~ NIST SP 800-12, An Introduction to Computer Security: The NIST Handbook
Auditing and Monitoring	17. Audit Trails	Audit and Accountability (AU)	~ NIST SP 800-12, An Introduction to Computer Security: The NIST Handbook
Computer Security Investigations and Forensics	14. Incident Response Capability	Incident Response (IR)	~ NIST SP 800-61, Computer Security Incident Handling Guide
Reviews, Inspections, Audits, and Other Evaluations Performed on Contractor Facilities and Operations		System and Services Acquisition (SA)	~ NIST SP 800-35, Guide to Information Technology Security Services
Component			
Configuration or Change Management Control	10. Hardware and Systems Software Maintenance 12. Documentation	Configuration Management (CM)	~ NIST SP 800-12, An Introduction to Computer Security: The NIST Handbook
Personnel Security	6. Personnel Security	Personnel Security (PS)	~ NIST SP 800-12, An Introduction to Computer Security: The NIST Handbook
Physical Security	7. Physical Security	Physical and Environmental Protection (PE)	~ NIST SP 800-12, An Introduction to Computer Security: The NIST Handbook

OMB A-11	NIST SP 800-26 Topic Area	NIST SP 800-53 Security Control Families	Implementation Guidance[24]
Operations Security	6. Personnel Security 7. Physical Security 8. Production, Input/Output Controls 9. Contingency Planning 10. Hardware and Systems Software 11. Data Integrity 12. Documentation 13. Security Awareness, Training, and Education 14. Incident Response Capability	System and Communications Protection (SC) Identification and Authentication (IA) Personnel Security (PS) Physical and Environmental Protection (PE) Incident Response (IR) System and Information Integrity (SI)	~ NIST SP 800-12, An Introduction to Computer Security: The NIST Handbook ~ NIST SP 800-26, Security Self-Assessment Guide for Information Technology Systems ~ NIST SP 800-53, Recommended Security Controls for Federal Information Systems
Program/System Evaluations Whose Primary Purpose is Other Than Security	2. Review of Security Controls 4. Authorize Processing	Audit and Accountability (AU)	~ NIST SP 800-12, An Introduction to Computer Security: The NIST Handbook ~ NIST SP 800-26, Security Self-Assessment Guide for Information Technology Systems ~ NIST SP 800-53, Recommended Security Controls for Federal Information Systems

www.ingramcontent.com/pod-product-compliance
Lightning Source LLC
Chambersburg PA
CBHW081853170526
45167CB00007B/2994